THE LIBRARY
ST. MARY'S COLLEGE OF MARYLAND
ST. MARY'S CITY, MARYLAND 20686

D1559309

Carbon-Functional Organosilicon Compounds

MODERN INORGANIC CHEMISTRY

Series Editor: John P. Fackler, Jr.
Texas A&M University

METAL INTERACTIONS WITH BORON CLUSTERS
Edited by Russell N. Grimes

HOMOGENEOUS CATALYSIS
WITH METAL PHOSPHINE COMPLEXES
Edited by Louis H. Pignolet

THE JAHN-TELLER EFFECT AND
VIBRONIC INTERACTIONS IN MODERN CHEMISTRY
I. B. Bersuker

MOSSBAUER SPECTROSCOPY
APPLIED TO INORGANIC CHEMISTRY
Edited by Gary J. Long

CARBON-FUNCTIONAL ORGANOSILICON COMPOUNDS
Edited by Václav Chvalovský and Jon M. Bellama

A Continuation Order Plan is available for this series. A continuation order will bring delivery of each new volume immediately upon publication. Volumes are billed only upon actual shipment. For further information please contact the publisher.

Carbon-Functional Organosilicon Compounds

Edited by
Václav Chvalovský
Institute of Chemical Process Fundamentals
Czechoslovak Academy of Sciences
Prague, Czechoslovakia

and
Jon M. Bellama
Department of Chemistry
University of Maryland
College Park, Maryland

Plenum Press •New York and London

Library of Congress Cataloging in Publication Data

Main entry under title:

Carbon-functional organosilicon compounds.

(Modern inorganic chemistry)
Includes bibliographical references and index.
1. Organosilicon compounds. I. Chvalovský, Václav. II. Bellama, Jon M., 1938–
III. Series.
QD412.S6C37 2984   547′.08   84-3438
ISBN 0-306-41671-9

©1984 Plenum Press, New York
A Division of Plenum Publishing Corporation
233 Spring Street, New York, N.Y. 10013

All rights reserved

No part of this book may be reproduced, stored in a retrieval system, or transmitted, in any form or by any means, electronic, mechanical, photocopying, microfilming, recording, or otherwise, without written permission from the Publisher

Printed in the United States of America

PREFACE

The term "carbon-functional organosilicon compound" is used for organosilicon compounds in which a functional group is bonded to an organic moiety that is in turn connected to silicon via a Si-C bond. Thus, only Si-C_n-Y compounds (Y designates a functional group) will be discussed in this book; Si-O-C_n-Y compounds will in general not be considered, although the latter group does include a large number of natural substances containing silylated hydroxyl groups. (Because of the differing importance of various Y groups, the reader will find some deviation from this restriction). Finally, compounds containing a silyl group as the functional group are not considered.

An overview of the field of organosilicon chemistry would show that in the last several decades the commercial synthesis of organosilicon products has increased substantially, both in annual production and also in the increasing variety of compounds produced. This increase in the number of commercially available carbon-functional monomers and polymers (silicone polymers) is most remarkable and is occurring because new applications are continually being found for these compounds.

As might be expected, the number of publications in this field is also increasing. The important position of silicon in the periodic table - between carbon, aluminum, and phosphorus - means that an understanding of the nature of the bonds in organosilicon compounds is quite important in order to understand the bonding in these other areas. In general, silicon compounds can be readily prepared, and most of them are relatively stable; thus, it is possible by comparing analogous organic and organosilicon compounds to draw some general conclusions concerning the effects of the comparative electropositivity, the greater atomic volume, the potential utilization of vacant orbitals, or the higher polarizability of silicon.

PREFACE

The purpose of this volume is to provide information about carbon-functional organosilicon compounds from four different viewpoints, each of which will be useful to industrial research and other scientific workers.

The first chapter considers the practical applications and uses of carbon-functional organosilicon compounds (1) in laboratory procedures of organic chemistry and biochemistry, and (2) in production of composite materials from plastics and resins containing inorganic fillers. Although carbon-functional silanes also have begun to find application as regulators of living processes, only basic data on the biological activity of silanes are reported in this chapter. This topic has been recently reviewed [1] in a special volume of the series "Topics in Current Chemistry".

The effects of intramolecular interactions on the chemical properties of carbon-functional organosilicon compounds are treated in Chapter 2. This general interpretation will be of interest to the reader interested in the utilization of silanes in synthesis.

Chapter 3 summarizes the effects of structure on the NMR spectral parameters of carbon-functional silanes and provides information about (1) the interpretive power of NMR in the study of the structure of these compounds, and (2) the nature of their intramolecular interactions. The first part of the chapter briefly considers ^{29}Si NMR spectroscopy and gives information that is required for a better understanding of this specialized field. For a more detailed treatment the reader is referred to recent reviews [2,3].

In Chapter 4, the goal is to discuss from a theoretical point of view the effect of structure on the properties of organosilicon compounds. The review does not duplicate the recent review by Kwart and King [4].

Finally, we would like to thank Professor R. Zahradnik (J. Heyrovsky Institute of Physical Chemistry and Electrochemistry of the Czechoslovak Academy of Sciences) for valuable comments and suggestions.

References

[1] Tacke R., Wannagat U., Voronkov M.G.: Bioactive Organo-Silicon Compounds. Berlin-Heidelberg-New York: Springer (1979).
[2] Schraml J., Bellama J.M.: ^{29}Si-Magnetic Resonance in the Determination of Organic Structures by Physical Methods, Vol. 6. New York: Academic Press (1978).
[3] Kintzinger J.P., Marsmann H.: Oxygen-17 and Silicon-29 in NMR-Basic Principles and Progress, Vol. 17. Berlin-Heidelberg-New York: Springer (1981).
[4] Kwart H., King K.G.: d-Orbitals in the Chemistry of Silicon, Phosphorus and Sulfur. Berlin-Heidelberg-New York: Springer (1977).

CONTENTS

1. SOME APPLICATIONS OF CARBON-FUNCTIONAL
     ORGANOSILICON COMPOUNDS
     Václav Chvalovský

1.1 INTRODUCTION   1

1.2 THE PRODUCTION OF CARBON-FUNCTIONAL
     ORGANOSILICON COMPOUNDS   2

1.3 USES AS DERIVATIZING AGENTS FOR NATURAL SUBSTANCES   4

1.4 USES AS STATIONARY PHASES FOR GAS-LIQUID
     CHROMATOGRAPHY   9

1.5 USES IN SURFACE TREATMENT OF INORGANIC MATERIALS   10

   1.5.1 As Coupling Agents for Inorganic Fillers
         of Polymers   11
   1.5.2 For Immobilization of Peptides and Proteins
         on Solid Carriers   18
   1.5.3 For Immobilization of Metal Complex
         Catalysts   25

1.6 REFERENCES   28

2. INTRAMOLECULAR INTERACTION IN THE CHEMICAL BEHAVIOR
     OF CARBON-FUNCTIONAL ORGANOSILICON COMPOUNDS
     Josef Pola

2.1 INTRODUCTION   35

2.2 SUBSTITUENT EFFECTS OF SILYL GROUPS IN ORGANIC
     REACTIONS   39

2.2.1	Reactions of Organometallic Compounds	39
2.2.2	Reactions of Alkenes	40
2.2.3	Reactions of Aromatic Compounds	46
2.2.4	Reactions of Compounds with Keto, Carboxyl, and Carbalkoxyl Groups	53
2.2.5	Reactions of Amines	55
2.2.6	Reactions of Alcohols, and Their Esters, Acetals, and Ethers	59
2.2.7	Reactions of Thiols and Sulfides	72
2.2.8	Reactions of Alkyl Halides	73

2.3 REACTIONS WITH PATHWAY DOMINATED BY THROUGH-SPACE INTERACTION BETWEEN SILICON AND A FUNCTIONAL GROUP 82

2.3.1	Reactions Involving 1,2-Interaction	83
2.3.2	Reactions Involving 1,3-Interaction	93
2.3.3	Reactions Involving 1,4-Interaction	98
2.3.4	Reactions Involving 1,5- or 1,6-Interaction	107

2.4 REFERENCES 111

3. NMR SPECTROSCOPY IN THE INVESTIGATION AND ANALYSIS OF CARBON-FUNCTIONAL ORGANOSILICON COMPOUNDS
Jan Schraml

3.1 INTRODUCTION 121

3.2 ^{29}Si NMR SPECTROSCOPY 123

3.2.1	Experimental Aspects	123
3.2.2	^{29}Si Chemical Shifts - Basic Facts	129
3.2.3	^{29}Si Shielding Theory and Model	132
3.2.4	^{29}Si Spin-Spin Coupling Constants	142

3.3 ALIPHATIC CARBON-FUNCTIONAL COMPOUNDS 142

3.3.1	^{1}H NMR Spectroscopy	143
3.3.2	^{13}C NMR Spectroscopy	153
3.3.3	^{29}Si NMR Spectroscopy	155
3.3.4	Typical Results	161

CONTENTS

3.4 AROMATIC CARBON-FUNCTIONAL COMPOUNDS 202

    3.4.1 NMR Spectroscopy of Functional Groups 203
    3.4.2 NMR Spectroscopy of Silyl Groups $SiX^1X^2X^3$ 212
    3.4.3 NMR Spectroscopy of the Connecting Chain 217

3.5 CONCLUSIONS 219

3.6 REFERENCES 220

4. THEORETICAL ASPECTS OF BONDING IN ORGANOSILICON CHEMISTRY
Robert Ponec

4.1 INTRODUCTION 233

4.2 VALENCE SHELL EXPANSION FROM THE POINT OF VIEW OF QUANTUM THEORY 234

    4.2.1 The Concept of Orbitals 234
    4.2.2 Classification of Atomic Orbitals 235
    4.2.3 Transformation Properties of d Orbitals 237
    4.2.4 d Orbitals and Hybridization 241
    4.2.5 Other Factors Influencing the Utilization of d Orbitals in Bonding 245
    4.2.6 The Variational Principle and d Orbital Participation 247

4.3 HYPERCONJUGATION 250

    4.3.1 General Introduction 250
    4.3.2 Hyperconjugation in Organosilicon Chemistry 250
    4.3.3 Theoretical Aspects of Hyperconjugation 254
    4.3.4 Hyperconjugation and the Accuracy of a Localized Description of Bonding 257

4.4 ELECTRONEGATIVITY 259

4.5 ELECTRONIC EFFECTS OF SILYL SUBSTITUENTS AND THE POSSIBILITIES OF THEIR CHARACTERIZATION 263

    4.5.1 Linear Free Energy Relationships 263

4.5.2 Fourier Component Analysis of Internal Rotation	267
4.5.3 The α-Effect and Acid-Base Properties	271
4.6 THE CHEMISTRY OF SILICENIUM IONS AND SILYL ANIONS	275
4.6.1 Silicenium Ions	275
4.6.2 Silyl Anions	277
4.7 COMMON ASPECTS OF CHEMICAL REACTIVITY OF CARBON-FUNCTIONAL ORGANOSILICON COMPOUNDS	277
4.8 THE CHEMISTRY OF MULTIPLE BONDED SILICON	282
4.9 CONCLUSIONS	288
4.10 ACKNOWLEDGMENT	291
4.11 REFERENCES	292
INDEX	299

ABBREVIATIONS

Ar   aryl

E   electrophile

Et   ethyl

M   metal

Me   methyl

Nu   nucleophile

Ph   phenyl

R   alkyl

R^X alkyl substituted by X in β-ω position to silicon

X   electronegative or unsaturated group

Y   halogen

Δ heat

1

SOME APPLICATIONS OF CARBON-FUNCTIONAL ORGANOSILICON COMPOUNDS

Václav Chvalovský

Institute of Chemical Process Fundamentals
Czechoslovak Academy of Sciences
Prague, Czechoslovakia

1.1 INTRODUCTION

The oldest and commonest types of silicone polymers are methyl-substituted and methylphenyl-substituted siloxanes. These materials have been utilized in particular because of their well-known thermal stability, hydrophobic character, nonpolarity, biological inactivity, and low chemical reactivity.

Organosilicon monomers were originally used only as hydrophobic agents. The development of carbon-functional organosilicon monomers and polymers has extended the range of these properties and thus also their applications in fields apparently unrelated to each other. However, there are two common features in the applications of properties of carbon-functional organosilicon compounds (carbon-functional silanes), compounds having a $X_3Si(CH_2)_mY$ formulation:

a. Hydrolysis of the reactive X substituents gives a siloxane, a trifunctional structure that shows not only a high affinity to glass, metal oxides, and similar inorganic materials, but also ensures the hardness of the surface.

b. The character of the Y functional group, its polarity, and its chemical reactivity determine the surface tension, the formation of chemical bonds with organic materials, and also the biological activity.

From consideration of such properties one can readily understand the common and proposed applications of carbon-functional organosilicon monomers as coupling agents for plastics and resins filled with inorganic materials (see Section 1.5.1), for immobilization of peptides and metal complex catalysts on inorganic supports (see Section 1.5.2 and 1.5.3), as agents for derivatization of photoelectrodes [1], as additives enhancing the resistance of asphalt mixtures for roadway coatings [2], as stabilizers of silicate water solutions [3], as a component of preparations resisting the shrinking of wool [4], as agents for formation of layers on polycarbonate lenses resistent to scratching [5], as alkali resistent glass fibers for reinforcing cement [6], as soil repellent coatings for automobile glass [7], and in numerous other applications. The most important and characteristic applications will be discussed in subsequent sections of this chapter.

New discoveries of biological activity of organosilicon compounds have attracted systematic attention to this area in the past fifteen years. It has been found, for example, that some of the (aminoalkyl) substituted silanes exhibit significant biological activity [8], while e.g. 3-mercatopropyl substituted derivatives are not distinctly toxic [9]. New aspects of the utilization of organosilicon compounds have been suggested by toxicological studies of low molecular weight methylphenyl substituted siloxanes that had been considered quite inert to biological processes; however, these compounds were found to influence the reproductive ability of experimental animals. This important aspect of the possible application of carbon-functional compounds has been recently reviewed [10].

1.2 THE PRODUCTION OF CARBON-FUNCTIONAL ORGANOSILICON COMPOUNDS

The direct synthesis of carbon-functional silanes from silicon and halogen derivatives, a process that is technologically well developed for the production of most of the important organosilicon monomers such as methylchloro-silanes [11, 13] and phenylchlorosilanes [12, 14], cannot be used for industrial production.

SOME APPLICATIONS

$$CH_3Cl + Si \xrightarrow{300°} (CH_3)_2SiCl_2 \text{ (+ limited amounts of} \quad (1)$$
$$(CH_3)_3SiCl, \text{ and other monomers)}$$

$$C_6H_5Cl + Si \xrightarrow[Cu]{500°} (C_6H_5)_2SiCl_2 + C_6H_5SiCl_3 \quad (2)$$

The direct synthesis proceeds only at high temperatures, at which temperature the higher alkyl halides undergo decomposition. This decomposition results in substantially lower yields of the direct syntheses of ethylchlorosilanes [15] and propylchlorosilanes [16], compared to the synthesis of methylchlorosilanes [11, 13]. However, because of their availability, methylchlorosilanes and phenylchlorosilanes are used as the starting material for synthesizing some carbon-functional derivatives.

$$C_6H_5SiCl_3 \xrightarrow[FeCl_3]{Cl_2} C_6H_2Cl_3SiCl_3 \qquad [17] \quad (3)$$

$$(CH_3)_3SiCl \xrightarrow[h\nu]{Cl_2} ClCH_2(CH_3)_2SiCl \qquad [18] \quad (4)$$

The Grignard synthesis, which is frequently used in the laboratory because of its versatility, is not suitable for commercial production. It is utilized only in special cases.

$$(CH_3)_2SiCl_2 + C_6F_5MgBr \longrightarrow C_6F_5(CH_3)_2SiCl \qquad [19] \quad (5)$$

The most frequently applied and technologically advantageous process for producing carbon-functional compounds is hydrosilylation.

$$Cl_3SiH + CH_2=CHCF_3 \xrightarrow{Pt} Cl_3SiCH_2CH_2CF_3 \qquad [20] \quad (6)$$

$$(C_2H_5O)_3SiH + CH_2=CHCH_2NH_2 \xrightarrow{H_2PtCl_6}$$
$$(C_2H_5O)_3SiCH_2CH_2CH_2NH_2 \qquad [21] \quad (7)$$

$$Cl_3SiH + CH_2=CHCH_2Cl \xrightarrow[H_2PtCl_6]{} Cl_3SiCH_2CH_2CH_2Cl \quad [22] \quad (8)$$

$$Cl_3SiH + CH_2=CHCN \xrightarrow[NR_3]{} Cl_3SiCH_2CH_2CN \quad [23] \quad (9)$$

Compounds containing functional groups bonded to the alkyl group in the α or the γ position to the silicon are relatively stable with respect to the action of polar agents, i.e., toward acid or base catalyzed cleavage of the Si-C bond; the β derivatives, however, easily undergo β-elimination [24].

$$\equiv SiCH_2CH_2Cl \xrightarrow{H_2O, NaOH} \equiv SiOH + CH_2=CH_2 + NaCl \quad (10)$$

For this reason, γ- rather than β-derivatives are used most frequently for commercial purposes (see Tables I and II).

1.3 USES AS DERIVATIZING AGENTS FOR NATURAL SUBSTANCES

Silylating agents are used extensively in the chemistry of natural products to replace the active hydrogen atoms of hydroxyl, amine, and sulfhydryl groups by silyl groups, especially the trimethylsilyl group [34]. In general, the silylated derivatives are more volatile and are also more thermally stable than the starting compounds [25], with both factors facilitating the separation of natural products by gas chromatography:

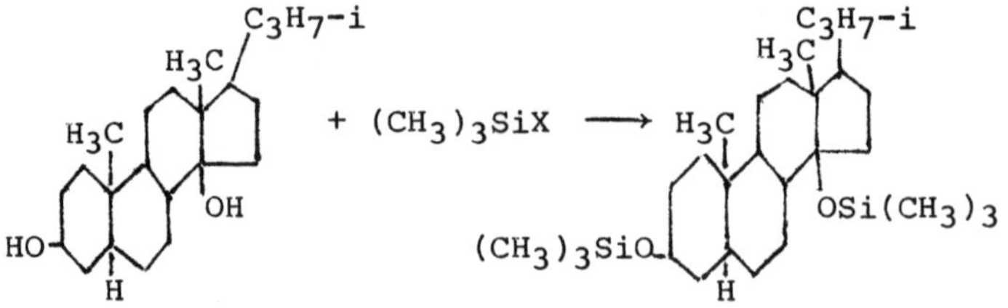

$$+ HX \quad (X=-Cl, -NHCCH_3, -NHSi(CH_3)_3, -N(CH_3)_2, \text{etc.}) \quad (11)$$

For certain purposes, derivatives of the trimethylsilyl group may be used instead of the trimethylsilyl group

Table I. Derivatizing agents

Compounds	Formula	Company
(chloromethyl)dimethylchlorosilane	$ClCH_2(CH_3)_2SiCl$	A,B,C
bis(chloromethyl)tetramethyldisilazane	$[ClCH_2(CH_3)_2Si]_2NH$	A
(bromomethyl)dimethylchlorosilane	$BrCH_2(CH_3)_2SiCl$	A,C
(pentafluorophenyl)dimethylchlorosilane	$C_6F_5(CH_3)_2SiCl$	A,B
(pentafluorophenyl)dimethylaminosilane	$C_6F_5(CH_3)_2SiNH_2$	B
(pentafluorophenyl)dimethyl-N,N-diethylamino-silane	$C_6F_5(CH_3)_2SiN(C_2H_5)_2$	B
allyldimethylchlorosilane	$CH_2=CHCH_2(CH_3)_2SiCl$	A

A - Fluka AG, B - Ventron GmbH, C - Applied Science Div., Milton Roy Co.

Table II. The most important carbon-functional organosilicon polymers used as stationary phases for gas-liquid chromatography

Type of Polymer	Upper Temperature Limit °C	Polarity	Solvent	Product Number	Company
methylvinylsiloxane	300	1	C,T	DC-430	Dow Corning
	300	1	C,T	GE SE-33	General Electric
	300	1	C,T	UC-W 982	Union Carbide
methylphenylvinylsiloxane	300	1	C,T	GE SE-54	General Electric
methyl(trichlorophenyl)siloxane	300	1	C,T	DC-560	Dow Corning
	300	1	C,T	GE Versilube F-50	General Electric
methyl(3,3,3-trifloropropyl)-siloxane	250	2	A,T	DC QF-1	Dow Corning
	275	2	A,T	OV-202 OV-210 OV-215	Ohio Valley Specialty Chemical Co.
	250	2	A,T	FS-1265	Merck

Table II continued

	200	2	A,T	UCON 50-HB-2000	Union Carbide
	250	2	A,T	FS-16	USSR
	250	2	A,T	FS-1	W.G. Pye & Co. Ltd.
methyl(3,3,3-trifluoro-propyl)vinylsiloxane	275	2	A,T	DC LSX-3-0295	Dow Corning
	275	2	A,T	Silastic 420	Dow Corning
methyl(2-cyanoethyl)siloxane	250	2	A	GE XE-60	General Electric
	200	3	A	GE XF-1150	General Electric
ethylenesuccinate(2-cyano-ethyl)siloxane copolymer	210	3	C,T	ECNSS-M	Pierce
	210	3	C,T	ECNSS-S	Pierce
methyl(3-cyanopropyl)siloxane	250	3	C,T	OV-105	Ohio Valley Specialty Chemical Co.
methylphenyl(3-cyanopropyl)-siloxane	250	2	C,T	OV-225	Ohio Valley Specialty Chemical Co.

continued

Table II continued

3-(cyanopropyl)siloxane	275	3	C,T	SILAR-10C Silar Laboratories Inc.
phenyl(cyanoalkyl)siloxane	275	3	C,T	SILAR-5CP, SILAR-7CP, SILAR-9CP Silar Laboratories Inc.
carboranemethylsiloxane	450	1	C,T	Dexsil 300 GC Dexsil Corp.
carboranemethylphenylsiloxane	450	1	C,T	Dexsil 400 GC Dexsil Corp.
carboranemethyl(cyanoethyl)siloxane	450	1	C,T	Dexsil 410 GC Dexsil Corp.

1 = non polar, 2 = medium polar, 3 = polar
A = acetone, C = chloroform, T = toluene

itself. For example, (chloromethyl)dimethylsilylation by commercial reagents (most frequently a mixture of (chloromethyl)dimethylchlorosilane and bis(chloromethyl)tetramethyldisilazane) [26, 27] is advantageous since (halogenomethyl)dimethylsilyl derivatives are highly sensitive to electron capture detection [28-30] because of the presence of the halogen. When compared to trimethylsilyl derivatives, the greater mass of the (chloromethyl)dimethylsilyl groups results in a two to threefold increase in retention time and also increases the differences between the retention times of polysilylated derivatives, e.g., carbohydrates. Increasing the retention times of the derivatives, of course, facilitates the separation of the mixtures of carbohydrates [31-33].

Halogen-containing silyl groups having a large mass, e.g., (bromomethyl)dimethylsilyl groups [32, 35] or (pentafluorophenyl)dimethylsilyl groups (flophemesyl) [36-38] are used to derivatize natural substances in cases where the mass spectrum of a given compound must be made more distinctive and the determination of its structure made easier (see the survey of the most common silylating agents in Table I). Flophemesyl-substituted derivatives, especially of alcohols and steroids, have the highest sensitivity of all of the silylated compounds [39] and can be determined quantitatively by electron capture detection. Silylation of steroids can also be effected advantageously by using allyldimethylchlorosilane. The silylated drivatives are comparatively stable to hydrolysis and have distinctive mass spectra [35,40,41].

1.4 USES AS STATIONARY PHASES FOR GAS-LIQUID CHROMATOGRAPHY

Organosilicon polymers fulfill very well the requirements for liquid stationary phases [42]. Depending on their structures, these polymers are good solvents for a variety of organic compounds and also can be good differential solvents for individual components of the sample. The polymers also can be produced with molecular weights sufficiently high that their vapor pressures do not exceed 10 Pa for a sufficiently long lifetime. The thermal stability of organosilicon polymers is markedly higher than most organic stationary substances, although it is affected somewhat by the structure of the carbon-functional substituents on the silicon. Moreover, the chemical inertness of the silicone phase is sufficient for the analysis of the majority of organic compounds.

The most widely used methyl substituted siloxanes (e.g., DC-200, GE SE-30, JXR, SP-2100, UC L-45) and methylphenyl substituted siloxanes (e.g., DC-704, DC-710, OV-3, GE SE-52, SP-2250) show maximal thermal stability at temperatures of 300° (the diphenyl substituted siloxanes up to 350°; the carboranemethyl substituted siloxanes up to 450°) [46, 47] and are nonpolar.

As can be seen from Table II, the polarity of siloxanes changes over a broad range according to the number of carbon-functional groups. A (cyanopropyl)siloxane liquid phase can easily resolve the pyrrolidine derivatives of carboxylic acids [48], polyunsaturated fatty acid methyl esters [46], phenylthiohydantoin amino acids [47], etc., and a 3,3,3-trifluoropropyl substituted siloxane proved to be highly effective for the separation of some optical isomers of carboxylic acids [48]. Thus, the versatility and thermal stability [49] of carbon-functional organosilicon polymers places them among the better GLC phases.

1.5 USES IN SURFACE TREATMENT OF INORGANIC MATERIALS

Organosilicon monomers are frequently used to modify the surfaces of various inorganic materials such as glass, silica, kieselguhr, and ceramics, and they are also used in the building industries for treatment of plasters and roofings. For these purposes the most frequently used compounds are monoorgano-substituted silanes of the $RSiX_3$ type in which X represents hydrolyzable groups such as the chloro, amino, alkoxy, or acyloxy group, according to the application. Alkoxy groups are hydrolyzed at the slowest rate; with the other groups, one should take into account that hydrolysis leads to the evolution of hydrogen chloride, carboxylic acid, or ammonia, the presence of which may be undersirable in some applications.

The interaction of the silanes with inorganic surfaces likely proceeds most frequently via hydrolysis of the polar groups of the silane by the water adsorbed on the hydrophilic surface. However, in the case of an agent deposited from a dilute solution onto the surface, reaction also certainly occurs from the moisture in the air. It is presumed that hydrolysis leads to the formation of an organosilanetriol

$$RSiX_3 + 3H_2O \longrightarrow RSi(OH)_3 + 3HX \qquad (12)$$

which condenses rapidly to give oligo-organosiloxanes that are bonded to the surface via hydrogen bonding or via chemical bonds, as shown in Eq. (10). Silylation makes the hydrophilic surface of inorganic materials hydrophobic; therefore, the wettability of the surface by organic compounds increases. In the case of composite materials in which a marked difference in the coefficients of thermal expansion (see Table III) causes the mechanical properties to deteriorate with changes in temperature, organosilicon interlayers aid in reducing the tension between the support and the resin [50]. As can be seen from Table IV, silylation of an inorganic surface makes the surface tension of the support more similar to that of the polymer and thus affects their interaction.

The modern uses of silylation to control the interaction of organic materials with inorganic supports also find wide application [51] in influencing the thixotropic properties of coatings and lubricant greases and thus offer the possibility of fundamental changes in the properties of inorganic chromatographic materials, e.g., in reverse-phase thin-layer chromatography. As a rule, however, in these applications a specific interaction does not take place between the functional groups of the silylating agent and the organic component of the composite material; thus, in this case the less expensive methyl substituted silanes are used more frequently. Therefore, in successive paragraphs we shall pay attention to the surface treatment of inorganic materials by such carbon-functional silanes that lead to unambigious chemical reactions with organic components.

1.5.1 As Coupling Agents for Inorganic Fillers of Polymers

Physico-mechanical properties of filled composite materials, i.e., reinforced plastics, laminates, and rubbers, are improved by treating the filler with coupling agents, shown in Table V. The properties are improved partially because of the increased wettability of the polymer surface [52-55].

Table III. Coefficients of thermal expansion of some organic polymers and glasses ($\cdot 10^6/°C$)

epoxy resins	45-65	quartz glass	0.6
polyester resins	55-90	Rasotherm	3.3
phenolics	60-80	Jenatherm	4.7
polystyrene	60-80	normal glass	8-9.5
polypropylene	100-200	NaCa glass	6-11
silicone	160-180	special glasses	up to 20

Table IV. Surface tension (mJ/m^2) of silane-treated glass and some polymers

Glass treated by		Polymer	
untreated glass	110	polyperfluoromethacrylate	10.6
untreated fused silica	78	polytetrafluoroethylene	18.6
CH$_3$Si(OCH$_3$)$_3$	22.5	polypropylene	20-24
CH$_2$=CHSi(OC$_2$H$_5$)$_3$	25	polyethylene	31
CH$_2$=C(CH$_3$)COO(CH$_2$)$_3$Si(OCH$_3$)$_3$	28	polystyrene	33
NCCH$_2$CH$_2$Si(OCH$_3$)$_3$	34	polymethylmethacrylate	39
NH$_2$(CH$_2$)$_3$NH(CH$_2$)$_3$Si(OCH$_3$)$_3$	33.5	polyvinylchloride	40
NH$_2$(CH$_2$)$_3$Si(OC$_2$H$_5$)$_3$	35	polyethylenterephthalate	43
$\overset{O}{\text{CH}_2\text{-CHCH}_2\text{O}}$(CH$_2$)$_3$Si(OCH$_3$)$_3$	38.5-42.5	nylon 6/6	46
Cl(CH$_2$)$_3$Si(OCH$_3$)$_3$	40.5		
HS(CH$_2$)$_3$Si(OCH$_3$)$_3$	41		

Table V. Some carbon-functional silanes produced on an industrial scale

Compounds	Formula	Product Number
(3-chloropropyl)trichlorosilane	$ClCH_2CH_2CH_2SiCl_3$	Z-6010[a]
(3-chloropropyl)trimethoxysilane	$ClCH_2CH_2CH_2Si(OCH_3)_3$	Z-6076[a]
vinyltrichlorosilane	$CH_2=CHSiCl_3$	Z-1226[b], A-150[b]
vinyltrimethoxysilane	$CH_2=CHSi(OCH_3)_3$	A-171[b]
vinyltriethoxysilane	$CH_2=CHSi(OC_2H_5)_3$	A-151[b], GKZ-12[c] NV-1107[d]
vinyltris(methoxyethoxy)silane	$CH_2=CHSi(OCH_2CH_2OCH_3)_3$	A-172[b]
vinyltriacetoxysilane	$CH_2=CHSi(OOCCH_3)_3$	A-188[b], Z-6075[a]
(3-methacryloxypropyl)trimethoxysilane	$CH_2=C(CH_3)COO(CH_2)_3Si(OCH_3)_3$	A-174[b], Z-6030[a]
allyltriethoxysilane	$CH_2=CHCH_2Si(OC_2H_5)_3$	GKZ-18[c]
methylvinyldichlorosilane	$(CH_2=CH)CH_3SiCl_2$	Z-1227[a]
vinyltris(t-butylperoxy)silane	$CH_2=CHSi[OOC(CH_3)_3]_3$	Y-5712[b]
(3-aminopropyl)triethoxysilane	$NH_2(CH_2)_3Si(OC_2H_5)_3$	A-1100[b], ADE-3[c] NVB-1114[d]
[3-(2-aminoethylamino)propyl]trimethoxysilane	$NH_2CH_2CH_2NH(CH_2)_3Si(OCH_3)_3$	A-1120[b], Z-6020[a]
vinyltriacetoxysilane	$CH_2=CHSi(OOCCH_3)_3$	A-188[b], Z-6075[a]
(3-methacryloxypropyl)trimethoxysilane	$CH_2=C(CH_3)COO(CH_2)_3Si(OCH_3)_3$	A-174[b], Z-6030[a]
allyltriethoxysilane	$CH_2=CHCH_2Si(OC_2H_5)_3$	GKZ-18[c]

SOME APPLICATIONS 15

Table V continued

Compound	Formula	Product Number
methylvinyldichlorosilane	$(CH_2=CH)CH_3SiCl_2$	Z-1227[a]
vinyltris(t-butylperoxy)silane	$CH_2=CHSi[OOC(CH_3)_3]_3$	Y-5712[b]
(3-aminopropyl)triethoxysilane	$NH_2(CH_2)_3Si(OC_2H_5)_3$	A-1100[b], ADE-3[c] NVB-1114[d]
[3-(2-aminoethylamino)propyl]-trimethoxysilane	$NH_2CH_2CH_2NH(CH_2)_3Si(OCH_3)_3$	A-1120[b], Z-6020[a]
[3-bis(2-hydroxyethyl)aminopropyl]-trimethoxysilane hydrochloride	$(HOCH_2CH_2)_2N(CH_2)_3Si(OC_2H_5)_3 \cdot HCl$	A-1111[b]
[3-(styrylaminoethyl)aminopropyl]-trimethoxysilane hydrochloride	$CH_2=CHC_6H_4NH(CH_2)_2NH(CH_2)_3Si(OCH_3)_3 \cdot HCl$	Z-6032[a]
(3-glycidoxypropyl)trimethoxysilane	$\overset{O}{\overset{\frown}{CH_2-CHCH_2}}O(CH_2)_3Si(OCH_3)_3$	Z-6040[a], A-187[b], NVB-1115[d]
[2-(3,4-epoxycyclohexyl)ethyl]-trimethcxysilane	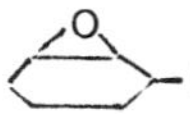 $\overset{O}{\overset{\frown}{\bigcirc}}-(CH_2)_2Si(OCH_3)_3$	A-186[b]
(3-mercaptopropyl)trimethoxysilane	$HS(CH_2)_3Si(OCH_3)_3$	Z-6062[a], A-189[b]

[a]Dow Corning, [b]Union Carbide, [c]USSR, [d]VEB Chemiewerk Nunchritz

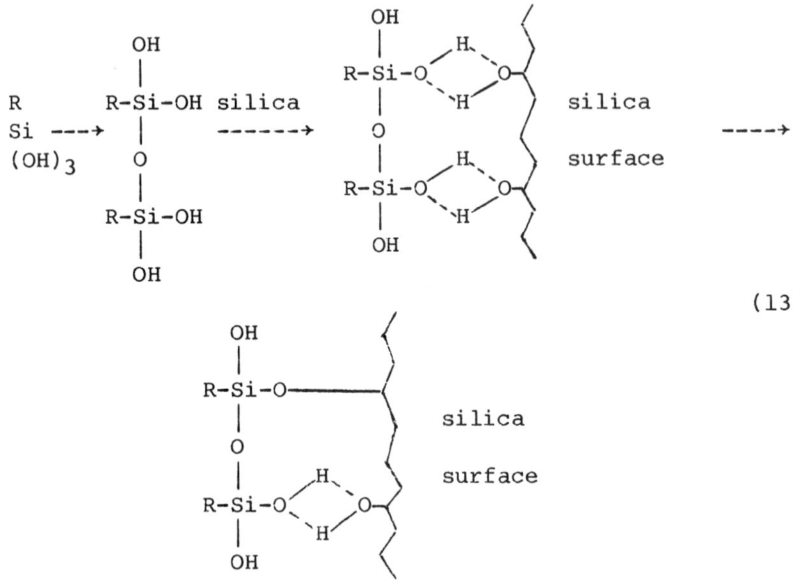

$$\underset{(OH)_3}{\underset{|}{R}} \xrightarrow{} \begin{array}{c} OH \\ | \\ R-Si-OH \\ | \\ O \\ | \\ R-Si-OH \\ | \\ OH \end{array} \xrightarrow{silica} \begin{array}{c} OH \\ | \\ R-Si-O\cdots H\cdots O \\ | \\ O \\ | \\ R-Si-O\cdots H\cdots O \\ | \\ OH \end{array} \Bigg\} \begin{array}{c} silica \\ surface \end{array} \xrightarrow{}$$

$$\begin{array}{c} OH \\ | \\ R-Si-O \\ | \\ O \\ | \\ R-Si-O\cdots H\cdots O \\ | \\ OH \end{array} \Bigg\} \begin{array}{c} silica \\ surface \end{array} \quad (13)$$

The most important of the physico-mechanical properties of the above composites is a result of the ability of the carbon-functional groups of the organosilicon agents (see Table V) to react with the functional groups of the polymers and to form covalent bonds between the polymer and the coupling agent [56, 57]. The great number of these reactive groups on the treated surface makes it possible to increase the network density of the composite [58, 59].

The functional group of a coupling agent (i.e., the chloropropyl substituent) can react at temperatures around 250° (via radical reactions not yet studied in detail) with comparatively unreactive thermoplastics such as polyethylene, polypropylene, or polystyrene [60]. Analogously, the (aminopropyl)silylated surface is capable of bonding to polyvinyl chloride, presumably by the addition of the amino group to the double bond formed by dehydrochlorination of the polymer at higher temperatures.

Chemically well defined reactions can also take place between coupling agents and thermosets containing reactive groups. For example, the phenolic hydroxyl group in phenolic resins reacts readily with the oxirane ring of epoxy substituted silanes.

SOME APPLICATIONS

$$\equiv SiCH_2CH_2\text{-[epoxy]} + \text{[Ph-CH}_2\text{-Ph-CH}_2\text{OCH}_2\text{-Ph]} \longrightarrow \text{[Ph-CH}_2\text{-Ph with HO-, -CH}_2\text{CH}_2\text{Si}\equiv, O\text{-substituents]} \quad (14)$$

In the case of epoxy resins, surface treatment with an amino-substituted silane is recommended [61], since this group also reacts with the oxirane ring under mild conditions.

$$\equiv Si(CH_2)_3NH_2 + CH_2\overset{O}{\underset{\diagdown}{C}}HCH_2(O\text{-}\langle Ph \rangle\text{-}\underset{CH_3}{\overset{CH_3}{C}}\text{-}\langle Ph \rangle\text{-}OCH_2\overset{OH}{C}HCH_2)_n \longrightarrow$$

$$\equiv Si(CH_2)_3NHCH_2\overset{OH}{C}HCH_2(O\text{-}\langle Ph \rangle\text{-}\underset{CH_3}{\overset{CH_3}{C}}\text{-}\langle Ph \rangle\text{-}OCH_2CHCH_2)_n \quad (15)$$

The free hydroxyl groups of epoxy resins and the functional groups of other polymers also react readily with the oxirane ring of coupling agents. Therefore, the surface treatment of inorganic fillers by (3-glycidyloxypropyl)trimethoxysilane improves the mechanical properties and the abrasion resistance of composites based on epoxy resins [62, 63], polyphenylene sulfide [64], and polyurethanes [65]. The coupling agent sometimes takes part in the crosslinking of the polymer itself.

In the case of unsaturated polyester resins that contain styrene, a bond is formed with an inorganic surface treated with vinylsilanes via radical copolymerization [66].

$$\equiv SiCH=CH_2 + \underset{\text{Ph}}{\overset{CH=CH_2}{|}} \longrightarrow \underset{\text{Ph}}{|}\underset{\overset{|}{Si}}{\overset{-CHCH_2CHCH_2CHCH_2-}{|}}\underset{\text{Ph}}{|} \quad (16)$$

Even better results are obtained if the unsaturated organosilicon agent is a 3-methacryloxypropyl substituted silane [67, 68], allylsilane [69], or other alkenylsilanes [70]. A similar difference in the reactivity of unsaturated silanes was also found in their application to the surface treatment of kaolin as the filler for polyethylene [71] or for ethylene-propylene-diene elastomer [70, 72]. In the case of a sulfur-vulcanized rubber, the more efficient agent turned out to be (3-mercaptopropyl)trimethoxysilane [57, 73, 74]. A comparative study of structure effects on the effectiveness of sulfur-containing agents for a styrene-butadiene rubber filled with silica showed (see Table VI) that the tensile strength and the elasticity modulus increase substantially on using silylalkyl substituted oligosulfides [75, 76].

Organosilicon agents are either initially introduced into powder fillers (treated fillers are produced commercially in some cases) or are used as one of the components in the production of filled plastics and rubbers. In the latter case carbon-functional silanes can act not only as a component of the polymer-agent-filler-agent-polymer system but also directly as the crosslinking component of the polymer-agent-polymer system.

Compared to the peroxide-initiated crosslinking of plastomeric polyolefins, the utilization of organosilicon compounds is of advantage because the Si-O-Si bonds can relax, which favorably influences the physico-mechanical properties of vulcanizates. The siloxane linkage between polymer chains can be formed either by the addition of vinyltrialkoxysilane to the double bond of the polymer [77, 78], followed by hydrolysis of the product formed, as shown in Eq. 17, or by the peroxide-initiated coupling of polymer chains by a siloxane that contains at least two unsaturated groups. Tetravinyltetramethylcyclotetrasiloxane $[(CH_2=CH)CH_3SiO]_4$ can be used as a suitable agent [79] for this purpose.

1.5.2 For Immobilization of Peptides and Proteins on Solid Carriers

Immobilization of Enzymes. Compared to homogeneous systems containing enzymes not bonded in cells, enzymes bonded to solid carriers show several advantages in their application in catalyzed reactions. They can be used in continuous processes without the problem of their separation products. Furthermore, the covalent bonding of the enzymes to suitable functional groups of the solid carrier also increases their stability to pH changes of the medium.

Table VI. Properties of silica-filled SBR with various experimental sulfur-containing silanes[a]

Organosilicon agent	M 200[b] MPa	Tensile strength MPa	Elongation %	Shore hardness °Sh
(no agent)	4.41	7.67	325	64
$(C_2H_5O)_3SiCH_2SH$	6.81	13.19	323	65
$[(C_2H_5O)_3SiCH_2]_2S$	5.26	9.01	310	64
$[(C_2H_5O)_3SiCH_2]_2S_2$	5.99	11.71	330	65
$[(C_2H_5O)_3SiCH_2]_2S_3$	5.92	10.06	300	64
$[(C_2H_5O)_3SiCH_2]_2S_4$	6.14	12.35	350	64
$(CH_3O)_3SiCH_2SH$	6.84	13.91	330	65
$[(CH_3O)_3SiCH_2]_2S_4$	5.87	11.89	340	65
$[(C_2H_5O)_3Si(CH_2)_3]_2S_3$	7.03	12.10	300	65
$[(C_2H_5O)_3Si(CH_2)_3]_2S_4$	6.89	11.10	280	65
$(CH_3O)_3Si(CH_2)_3SH$	7.87	14.74	290	65

[a] sulfur vulcanized at 140°C; [b] 200% modulus

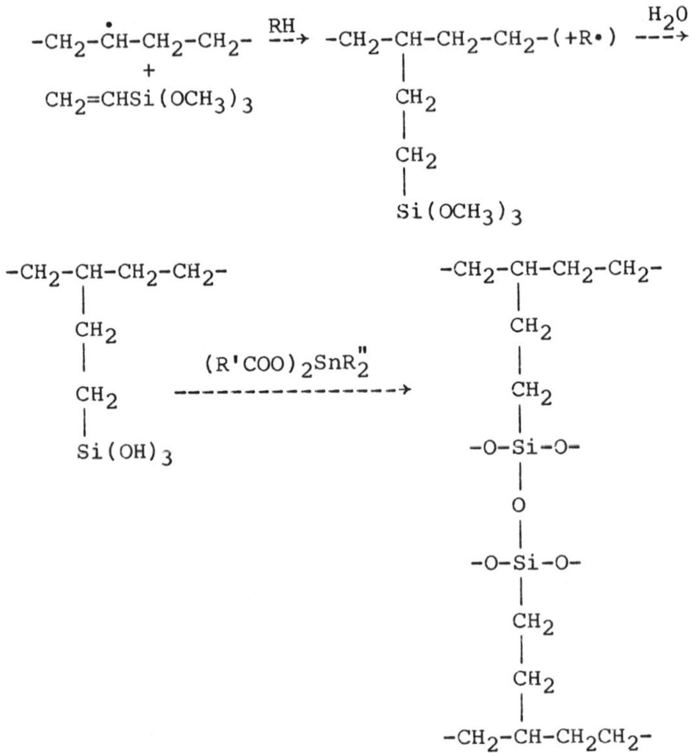

$$-CH_2-\overset{\cdot}{C}H-CH_2-CH_2- \;+\; CH_2=CHSi(OCH_3)_3 \xrightarrow{RH} -CH_2-CH(-CH_2-CH_2-CH_2-Si(OCH_3)_3)-CH_2-CH_2-(+R\cdot) \xrightarrow{H_2O}$$

$$-CH_2-CH(-CH_2-CH_2-Si(OH)_3)-CH_2-CH_2- \xrightarrow{(R'COO)_2SnR''_2} -CH_2-CH(-CH_2-CH_2-O-Si(-O-)-O-Si(-O-)-CH_2-CH_2-CH(-)-CH_2CH_2-)-CH_2-CH_2- \quad (17)$$

Organic polymers containing suitable functional groups or derivatized porous glass, silica, or ceramics can be used as the carrier. Inorganic carriers are advantageous, especially because of their greater compression resistance and stability of surface texture to changes in the reaction medium [80-84].

Derivatizing an inorganic surface by silanes from an aqueous medium usually leads to formation of a layer with a lower concentration of functional groups but with a longer lifetime than in the case of treatment by silanes dissolved in organic solvents.

The most frequently used coupling agents are (3-aminopropyl)trimethoxysilane and [3-(2-aminoethyl)aminopropyl]trimethoxysilane. The surface modified by these silanes is then treated with other agents that can react with the amino group of the silane to form derivatives capable of reacting with certain amino acids of the pep-

SOME APPLICATIONS

tides to be immobilized. Thus, for example, the reaction with glutaraldehyde leads to a derivative that can bond peptides via its aldehyde group.

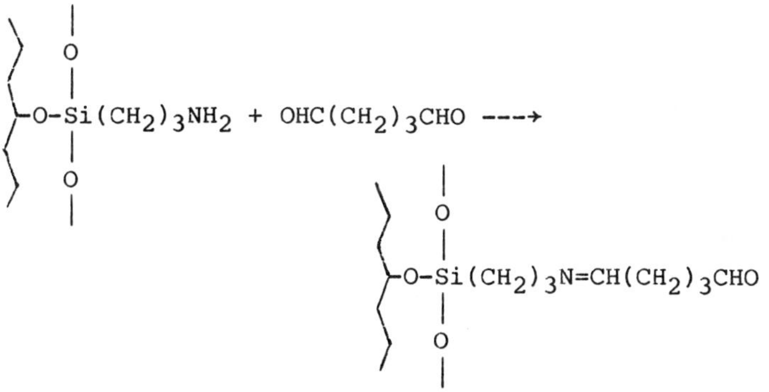

$$\rbrace\text{-O-Si(CH}_2)_3\text{NH}_2 + \text{OHC(CH}_2)_3\text{CHO} \longrightarrow \rbrace\text{-O-Si(CH}_2)_3\text{N=CH(CH}_2)_3\text{CHO} \tag{18}$$

The reaction with thiophosgene gives a derivative containing an isothiocyanato group [85, 86] that reacts easily with lysine residues in a natural medium.

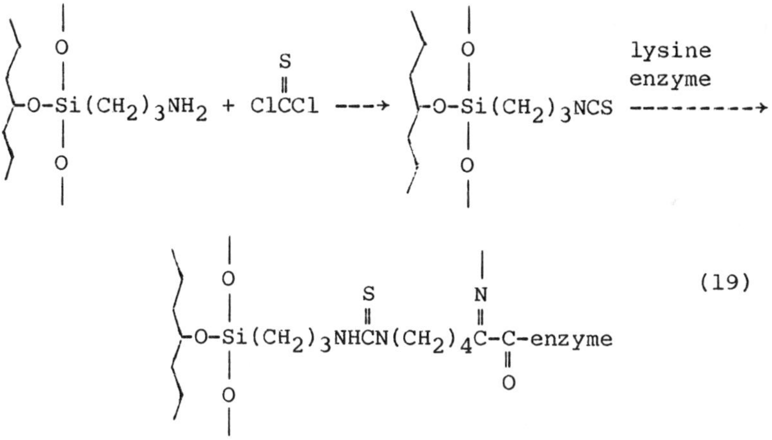

$$\rbrace\text{-O-Si(CH}_2)_3\text{NH}_2 + \text{ClCCl} \longrightarrow \rbrace\text{-O-Si(CH}_2)_3\text{NCS} \xrightarrow{\text{lysine enzyme}}$$

$$\rbrace\text{-O-Si(CH}_2)_3\text{NHCN(CH}_2)_4\text{C-C-enzyme} \tag{19}$$

(with S=C-NH and N=C linkages, C=O)

The reaction of the amino group of the silyl derivative with a carboxylic group in the presence of dicyclohexylcarbodiimide is especially convenient for enzymes that are to be coupled in acidic media [87], as in the case of pepsin [Eq. 20]. p-Nitrobenzoyl chloride can be used to substitute the amino group by a p-nitrobenzoyl group that, after reduction to the p-amino derivative [88] and conversion into the diazonium salt, can be bonded with advantage via an azo-linkage to the enzyme containing tyrosine [Eq. 21].

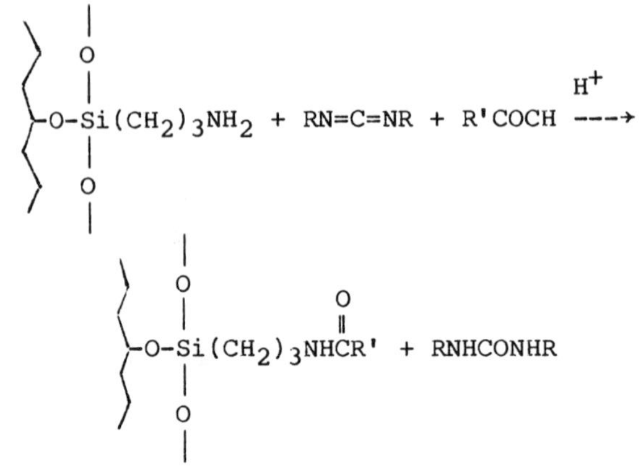

$$\begin{array}{c} \}\text{-O-Si(CH}_2)_3\text{NH}_2 + \text{RN=C=NR} + \text{R'COCH} \xrightarrow{\text{H}^+} \\ \\ \}\text{-O-Si(CH}_2)_3\text{NHCR'} + \text{RNHCONHR} \end{array}$$

(20)

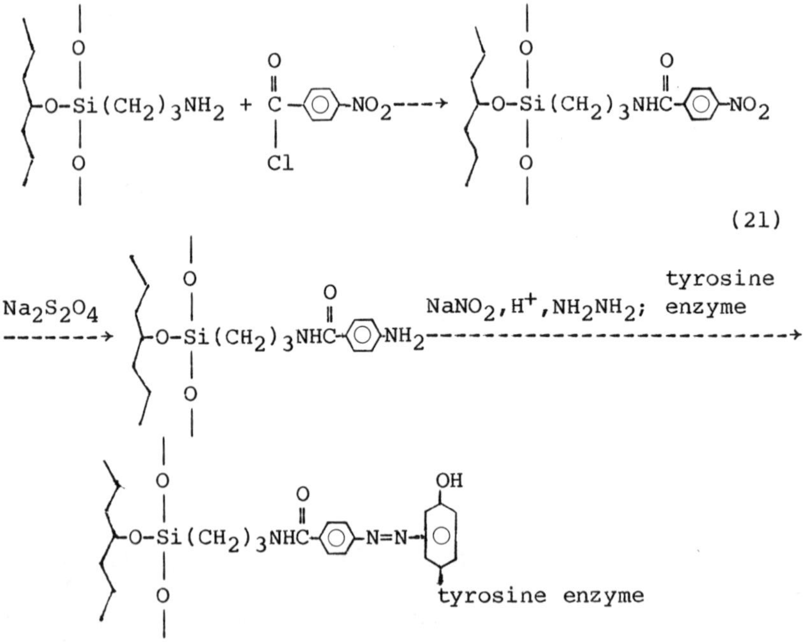

$$\}\text{-O-Si(CH}_2)_3\text{NH}_2 + \underset{\text{Cl}}{\overset{\text{O}}{\text{C}}}\text{-}\langle\text{O}\rangle\text{-NO}_2 \longrightarrow \}\text{-O-Si(CH}_2)_3\text{NHC-}\langle\text{O}\rangle\text{-NO}_2$$

(21)

$$\xrightarrow{\text{Na}_2\text{S}_2\text{O}_4} \}\text{-O-Si(CH}_2)_3\text{NHC-}\langle\text{O}\rangle\text{-NH}_2 \xrightarrow[\text{enzyme}]{\text{NaNO}_2,\text{H}^+,\text{NH}_2\text{NH}_2;\ \text{tyrosine}}$$

$$\}\text{-O-Si(CH}_2)_3\text{NHC-}\langle\text{O}\rangle\text{-N=N-}\langle\text{O}\rangle\text{-OH}$$
tyrosine enzyme

In addition to (3-aminopropyl)triethoxysilane, another agent used to bond the cysteine component of enzymes is (3-mercaptopropyl)triethoxysilane.

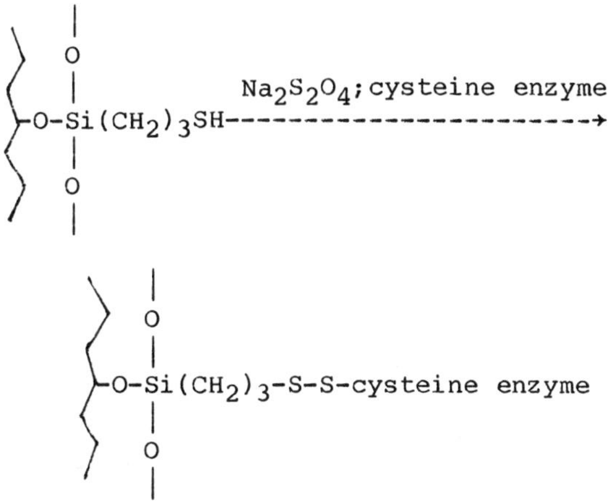

$$\left.\begin{array}{c}\\\\\\\\\end{array}\right\}\begin{array}{c}|\\O\\|\\-O-Si(CH_2)_3SH\\|\\O\\|\end{array}\xrightarrow{\text{Na}_2\text{S}_2\text{O}_4;\text{cysteine enzyme}}$$

(22)

$$\left.\begin{array}{c}\\\\\\\\\end{array}\right\}\begin{array}{c}|\\O\\|\\-O-Si(CH_2)_3-S-S-\text{cysteine enzyme}\\|\\O\\|\end{array}$$

A less frequently used coupling agent is 3-chloropropyltriethoxysilane.

The immobilization of enzymes has been applied on an industrial scale, e.g., in the hydrolysis of starch to glucose and in sugar inversion. For that reason the methods for immobilization on an industrial scale and the regeneration of enzymes are already well established [89-91].

Affinity Chromatography. Both non-polar alkylsilanes and polar carbon-functional silanes provide bonded phases for gas and liquid chromatography [92, 105]. The functional groups can be further modified for special purposes. For example, complexes of copper ions with (aminopropyl)silylated surfaces are able to adsorb the lower olefins selectively.

The most interesting applications of carbon-functional silanes in chromatographic methods were developed in affinity chromatography.

Inorganic granular support matrices for affinity chromatography, such as porous glass or silica, are advantageous because they are resistant to acids, to organic solvents, and to microbial attack. They are rigid, chemically and thermally stable, and they have outstanding hydrodynamic properties. A serious disadvantagae is their limited binding capacity and the non-specific adsorption or denaturation of proteins due to Si-OH groups on the silica surface. This deficiency can be eliminated by derivatization of the surface, e.g., by means of (3-glycidoxypropyl)trimethoxysilane [93-95] and hydrolysis that produces (3-glycerylpropyl)silylated silica. These packings have been used for high performance gel permeation chromatography of proteins [96-98, 107], water soluble cellulose derivatives [99, 104], etc. In the industrial purification by affinity chromatography of antithrombin and heparin, the hydroxyl groups of the (3-glycerylpropyl)siloxy silica can be etherified with the methylol groups of N-methylolacrylamide co-polymer [100]. Glycyl-D-phenylalanine immobilized by attachment with glutaraldehyde to (aminopropyl)silylated glass was found to purify efficiently the carboxypeptidase [101]. Other chiral bonded phases consist of attaching L-proline to the (aminopropyl)silylated glass [102] or in grafting, e.g., L-valine-tert-butylamide on to (cyanopropyl)methylphenylmethylsiloxane (OV-225; see Table II). These chiral phases can be used for the resolution of the amino acid enantiomers [103].

Solid State Synthesis and Sequencing of Peptides and Proteins. In the synthesis of the larger peptides, the first and most frequently used carriers were organic polymers, especially a chloromethylated styrenedivinylbenzene co-polymer (Merrifield Resin). This carrier was then replaced by porous glass beads or by silica derivatized with (chloromethyl)phenylalkylsilanes and further treated with triethylenetetramine derivatives. The reaction of the carrier prepared as above with the triethylammonium salt of the protected amino acid yields the relatively stable ester to which other amino acids can be attached by using common methods of peptide synthesis. The solid state synthesis is advantageous particularly because of the possibility of automation that facilitates the preparation of the larger peptides [106]. The method based on a (chloromethyl)phenylalkylsilylated carrier is also superior to that based on the Merrifield Resin because it prevents the formation of carcinogenic intermediate products. Furthermore, the inorganic carrier is cheaper and more stable.

The attachment of proteins and larger peptides via certain groups of a given amino acid to the derivatized

surface of inorganic carriers is also used to determine
their structures (the sequence of their amino acids) by
Edman degradation. The inorganic surface is silylated most
frequently by (3-aminopropyl)triethoxysilane (A) or by
[3-(2-aminoethyl)aminopropyl]trimethoxysilane (B). The
treated carriers can be used to bond a peptide either
through its end carboxylic group using dicyclohexylcarbodiimide (see Eq. 20), or alternatively the derivatized
carrier is treated with p-phenylenediisothiocyanate solution [107, 108].

In these reactions the surfaces treated with compound
(A) are less reactive than surfaces treated with derivative
(B). The latter compounds are recommended for the attach-
ment of layer peptides [109].

Edman degradation of the immobilized peptides can
then be effected by a number of established automated
methods [110, 111].

1.5.3 For Immobilization of Metal Complex Catalysts

Metal complex catalysts acting in a homogeneous
liquid phase frequently show higher catalytic activity,
selectivity, and stereospecificity, and they operate at
lower reaction temperatures than solid heterogeneous catalysts. On the other hand, their disadvantage is the frequently difficult and uneconomic separation from the
reaction products. This problem can be solved in a
fashion similar to that of the enzymes cited above by
immobilization of the metal complex catalysts on organic
polymers or inorganic supports, most frequently on silica.
Inorganic supports are again advantageous because they
have a constant texture, and their porosity is not changed
by the action of solvents, pressure, and temperature.

Metal complexes can be bonded to inorganic supports
by adsorption [112, 113] or by specific reactions of the
complexes with the surface hydroxyl groups of the support
[114, 115]. The widest application is that of spacers
using carbon-functional silanes containing functional
groups acting as ligands of the metal complex. For this
purpose the most frequently used compounds are alkoxy
substituted silylalkylphosphanes that can be prepared,
e.g., by the addition of diphenylphosphane to vinyl

triethoxysilane [116].

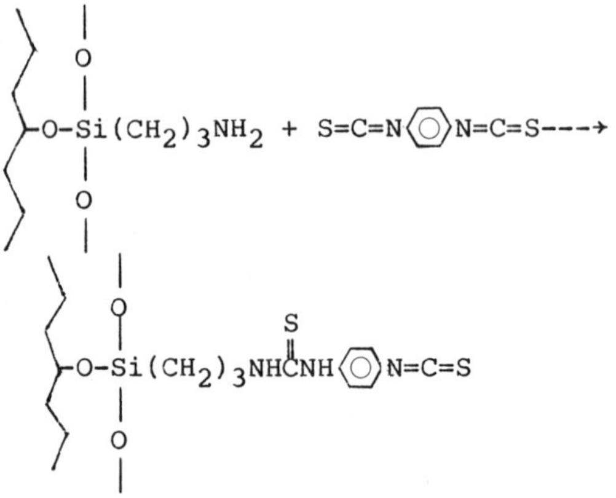

$$\}\text{-O-Si(CH}_2)_3\text{NH}_2 + \text{S=C=N}\langle\bigcirc\rangle\text{N=C=S} \dashrightarrow$$

(23)

$$\}\text{-O-Si(CH}_2)_3\text{NH}\overset{\text{S}}{\overset{\|}{\text{C}}}\text{NH}\langle\bigcirc\rangle\text{N=C=S}$$

The reaction with the amino group of the organosilicon coupling agent produces a derivative in which the free isothiocyanate group reacts with the amine group of a protein

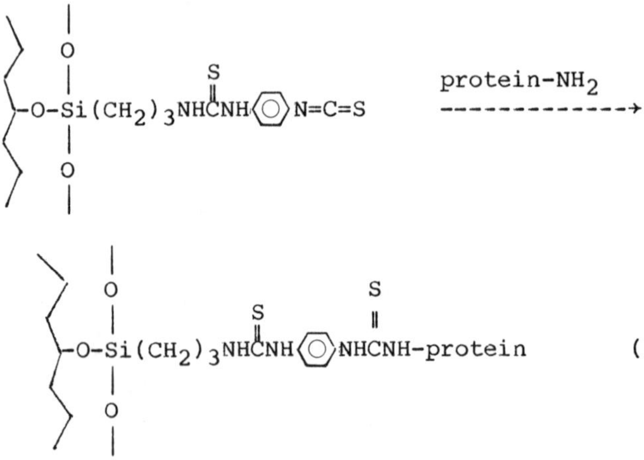

$$\}\text{-O-Si(CH}_2)_3\text{NH}\overset{\text{S}}{\overset{\|}{\text{C}}}\text{NH}\langle\bigcirc\rangle\text{N=C=S} \quad \xrightarrow{\text{protein-NH}_2}$$

$$\}\text{-O-Si(CH}_2)_3\text{NH}\overset{\text{S}}{\overset{\|}{\text{C}}}\text{NH}\langle\bigcirc\rangle\text{NH}\overset{\text{S}}{\overset{\|}{\text{C}}}\text{NH-protein} \quad (24)$$

$(C_2H_5O)_3SiCH=CH_2 + HP(C_6H_5)_2 \rightarrow (C_2H_5O)_3SiCH_2CH_2P(C_6H_5)_2$ (25)

or by lithium synthesis

$(C_2H_5O)_3SiCH_2Cl + LiP(C_6H_5)_2 \dashrightarrow (C_2H_5O)_3SiCH_2P(C_6H_5)_2$ (26)

from (halogenoalkyl)trialkoxysilanes. Surface treatment of

SOME APPLICATIONS   27

a silica surface with silylalkylphosphanes leads to an
immobilization of the phosphanes to which a suitable metal
complex can then be bonded.

$$(C_2H_5O)_3SiCH_2CH_2P(C_6H_5)_2 \xrightarrow{\text{silica}} \begin{Bmatrix} \\ \\ \\ \end{Bmatrix} \begin{matrix} | \\ O \\ | \\ -O-SiCH_2CH_2P(C_6H_5)_2 \\ | \\ O \\ | \end{matrix} \quad (27)$$

An alternative procedure is the synthesis of metal complexes containing alkoxysilylalkyl substituted phosphane
ligands and modification of the support surface by this
agent [117]. Another alternative of surface modification
is also its silylation by (chloropropyl)trichlorosilane and
the amination of the chloropropyl group, followed by coordination of the metal complex via the amino group [118].
Coupling agents are sometimes synthesized on a silica surface by the functionalization of the unsubstituted polymer.
Thus, for example, hydrolysis of phenyltrichlorosilane
gave octaphenylsilsesquioxane that was then deposited on a
silica surface. Its phenyl groups were chloromethylated
and substituted with diphenylphosphane groups, using a
lithium condensation [119] as shown in Eq. 28.

$$C_6H_5SiCl_3 \xrightarrow{H_2O} (C_6H_5SiO_{1.5})_8 \xrightarrow{\text{silica}}$$

$$\begin{Bmatrix} \\ \\ \\ \end{Bmatrix} \begin{matrix} | \\ O \\ | \\ -O-SiC_6H_5 \\ | \\ O \\ | \end{matrix} \xrightarrow[ZnCl_2]{CH_3OCH_2Cl} \begin{Bmatrix} \\ \\ \\ \end{Bmatrix} \begin{matrix} | \\ O \\ | \\ -O-Si-C_6H_4CH_2Cl \\ | \\ O \\ | \end{matrix} \xrightarrow{LiP(C_6H_5)_2}$$

$$\begin{Bmatrix} \\ \\ \\ \end{Bmatrix} \begin{matrix} | \\ O \\ | \\ -O-SiC_6H_4CH_2P(C_6H_5)_2 \\ | \\ O \\ | \end{matrix} \quad (28)$$

The problem of reactions taking place during immobilization of metal complexes with the aid of organosilicon spacers has already attracted much attention [120-126]; the effect of the length of the aliphatic chain of coupling agents of the $\equiv Si(CH_2)_m P(C_6H_5)_2$ type on the catalytic activity of immobilized rhodium [I] complexes in hydrogenation [117] and hydrosilylation [130] of 1-alkenes has been studied in detail. The immobilized rhodium [I] complexes containing chiral ligands have been used with success in enantioselective hydrosilylation and hydrogenation [128].

The immobilization of metal complexes has been recently surveyed in reviews [129] and monographs [130].

1.6 REFERENCES

1. Bocarsly A.B., Walton E.G., Bradley M.G., Wrighton M.S.: J. Electroanal. Chem. Interfacial Electrochem. 100, 283 (1979).
2. Crawford W.C., Wilson J.R.: US 4,173.489 (1979).
3. Plueddemann E.P.: Ger. Offen. 2,912.430 (1979).
4. Koerner G., Schmidt G., Nickel F.: Ger. Offen.
6. Otomo Koichiro, Yosthimura Takuji, Kudo Shozo: Jpn. Kokai Tokkyo Kogo 79,116,029 (1979).
7. Inoue Yoshio, Hidaka Ryukaro, Nakamura Katsui, Inosaka Toshifumi: Jpn. Kokai Tokkyo Kogo 79,118,404 (1979).
8. Parent R.A.: Drug Chem. Toxicol. 2(3), 295 (1979).
9. Lukevics E.: Biological Activity of Nitrogen-Containing Organosilicon Compounds. In: Biochemistry of Silicon and Related Problems. Plenum Press, New York (1978).
10. Tacke R., Wannagat U., Voronkov M.G.: Bioactive Organo-Silicon Compounds. Berlin-Heidelberg-New York: Springer (1979).
11. Joklík J., Bažant V.: Collect. Czech. Chem. Commun. 26, 417 (1961).
12. Andrianov K.A., Golubtsov S.A., Tishina N.N., Trofimova I.V.: Zhur. Prikl. Khim. 32, 201 (1959).
13. Rochow E.G.: J. Amer. Chem. Soc. 67, 963 (1945).
14. Rochow E.G.; Gilliam W.F.: J. Amer. Chem. Soc. 67, 1772 (1945).
15. Meals R.N.: J. Amer. Chem. Soc. 68, 1880 (1946).

16. Nametkin N.S., Topchiev A.V., Kartasheva L.I.: Dokl. Akad. Nauk SSSR 101, 885 (1955).
17. Yakubovich A.Ya., Motsarev G.V.: Dokl. Akad. Nauk SSSR 99, 1015 (1954).
18. Runge F., Zimmermann W.: Chem. Ber. 87, 282 (1954).
19. Whitingham A., Jarvie A.W.P.: J. Organometal. Chem. 13, 125 (1968).
20. Tarrant P., Dyckes G.W., Dunmire R., Butler G.B.: J. Amer. Chem. Soc. 79, 6536 (1957).
21. Petrov A.D., Ponomarenko V.A. Sokolov B.A., Odabashyan G.V.: Izv. Akad. Nauk SSSR, Otd. Khim. Nauk 1957, 1206.
22. Fialová V., Bažant V., Chvalovský V.: Collect. Czech. Chem. Commun. 38, 3837 (1973).
23. Saam J.C., Speier J.L.: J. Org. Chem. 24, 427 (1959).
24. Sommer L.H., Dorfman E., Goldberg G.M., Whitmore F.C.: J. Amer. Chem. Soc. 68, 488 (1946).
25. Ikekawa N., Hattori F., Rubio-Lightbourn J., Miyzaki H., et al.: J. Chromatogr. Sci. 10, 233 (1972).
26. Morita H., Montgomery W.G.: J. Chromatogr. 123, 454 (1976).
27. Harvey D.J.: J. Chromatogr. 147, 291 (1978).
28. Thomas B.S., Eaborn C., Walton D.R.M.: Chem. Commun. 408 (1966).
29. Eaborn C., Holder C.A., Walton D.R.M., Thomas B.S.: J. Chem. Soc. (C) 2502 (1969).
30. Harvey D.J., Paton W.D.M.: J. Chromatogr. 109, 73 (1975).
31. Brooks C.J.W., Middleditch B.S.: Anal. Lett. 5, 611 (1972).
32. Chapman J.R., Bailey E.: Anal. Chem. 45, 1636 (1973).
33. Chapman J.R., Bailey E.: J. Chromatogr. 89, 215 (1974).
34. Brittain G.D., Schewe L.: Recent Advances in Gas Chromatography (Domsky I.I., Perry J.A., Ed.), M. Dekker Inc., New York, (1971).
35. Poole C.F., Zlatkis A.: J. Chromatogr. Sci. 17, 115 (1979).
36. Morgan E.D., Poole C.F.: J. Chromatogr. 89, 225 (1974).
37. Poole C.F., Morgan E.D.: Org. Mass Spectrom. 10, 537 (1975).
38. Francis A.J., Morgan E.D., Poole C.F.: Org. Mass Spectrom. 11, 671 (1978).
39. Francis A.J., Morgan E.D., Poole C.F.: J. Chromatogr. 161, 111 (1978).

40. Phillipou G.: J. Chromatogr. 129, 384 (1976).
41. Poole C.F., Singhawangcha S., Hu L.G. Chan, et al.: J. Chromatogr. 187, 331 (1980).
42. McNair H.M., Bonnelli E.I.: Basic Gas Chromatography, Varian Aerograph, Walnut Creek, California (1969).
43. Andrianov K.A., Izmailov B.A., Kalinin V.N., Zakharkin L.I., et al.: USSR 690.021 (1979).
44. Andrianov K.A., Izmailov B.A., Kalinin V.N., Zakharkin L.I., et al.: USSR 690.019 (1979).
45. Satouchi Kiyoshi, Saito Kunihiko: Biomed. Mass Spectrom. 6, 144 (1979).
46. Jamieson G.R., McMinn A.L., Reid E.H.: J. Chromatogr. 178, 555 (1979).
47. Johnson N.D., Hunkapiller M.W., Hood L.E.: Anal. Biochem. 100, 335 (1979).
48. Horiba Masao, Kitahara Hajimu, Takahashi Kenichi, Yamamoto Seiya et al.: Agric. Biol. Chem. 43, 2311 (1979).
49. Pavlova V.B., Metkin I.A., Nikiforova G.N.: Khimiya i Praktich. Primenenie Kremnii- i Fosfororganich. Soedin. 1979, 37.
50. Zisman W.A.: Ind. Eng. Chem., Prod. Res. Dev. 8, 98 (1969).
51. Lee L.H.: Soc. Plast. Ind. RPC Proc. 23, 9D (1968).
52. Plueddemann E.P.: J. Adhes. 2, 184 (1970).
53. Bascon W.B., Romans J.B.: Ind. Eng. Chem. Prod. Res. Develop. 7, 172 (1963).
54. Jelínek P., Mikšovský F.: Sb. Ved. Pr. Vys. Sk. Báňské v Ostravě, Řada Hut. 23, 201 (1977).
55. Sawai Michio: Jpn. Kokkai Tokkyo Kogo 79, 96,192 (1979).
56. Cameron G.M., Ranney M.W., Sollmann K.J.: Europ. Rubber J. 156, 37 (1974).
57. Zemianski L.P., Pagano C.A., Ranney M.W.: Rubber World 196, 53 (1970).
58. Wagner E., Brünner H.: Angew. Chemie 72, 744 (1960).
59. Wagner M.P.: Rubber Chem. Technol. 49, 703 (1976).
60. Plueddemann E.P.: Polymer-Plast. Technol. Eng. 2, 89 (1973).
61. Losev V.B., Aslanova M.S., Voitsekhovich N.Ya., Meitin Yu.V., Fridman G.Ye.: Plast. Massy 1967 (10), 61.
62. Leinen R.W., Pieterick J.A., Platcher W.A., Robins J.: US 4,168.332 (1979).
63. Tanaka Goro, Suzuki Hiroshi, Uzano Takeshi: Jpn. Kokai Tokkyo Koho 79,110,298 (1979).

SOME APPLICATIONS

64. Needham D.G.: US 4,176.098 (1979).
65. Mizuno Shieji, Adachi Tsuneyuki: Jpan. Kokai Tokkyo Koho 79,132,699 (1979).
66. Wagner G.H., Bailey D.L., Pines A.N., Dunham M.L., McIntyre D.B.: Ind. Eng. Chem. 45, 367 (1953).
67. Cassidy P.E., Yager B.J.: J. Macromol. Sci. D1, 1 (1971).
68. Naketsuka Takuo, Kawasaki Hitoshi, Itadeni Katsuhiko, Yamashika Shinzo: J. Appl. Polym. Sci. 24, 1985 (1979).
69. Mikhalskii A.I.: Usp. Khim. 29, 2050 (1970).
70. Vondráček P., Schätz M.: Elektroisol. Kablová Tech. 32, 17 (1979).
71. Lipatov Yu.S., Gede I., Svyatenko G.P., et al.: Kompozitsion. Polymer. Materialy, Kiev 1979, (2), 8.
72. Ranney M.W., Pagano C.A.: Rubber Chem. Technol. 44, 1080 (1971).
73. Grillo T.A.: Rubber Age 103 (8) 37.
74. Grillo T.A.: US 3,834.924 (1974).
75. Vondráček P., Hradec M., Chvalovský, V., Hua Dang Khanh: J. Appl. Pol. Sci. (in press).
76. Wolff S.: Kautch., Gummi, Kunstst. 32 (10), 760 (1979).
77. Santelli T.R.: US 3,075.948 (1963).
78. Scott H.G.: US 3,646.155 (1972).
79. MacKenzie B.T. Jr.: US 3,859.247 (1975).
80. Wetall H.H.: Science 166, 615 (1969).
81. Wetall H.H.: Nature (London) 223, 959 (1969).
82. Melrose G.J.H.: Rev. Pure Appl. Chem. 21, 83 (1971).
83. Zaborsky O.R.: Immobilized Enzymes, CRC Publishing, Cleveland, Ohio 1974.
84. Messing R.A., Stinson H.R.: Mol. Cell. Biochem. 4, 217 (1974).
85. Wachter E., Hofner H., Otto J.: FEBS Lett. 35, 97 (1973).
86. Machleidt W., Wachter E., Scheulen M., Otto J.: FEBS Lett. 36, 217 (1973).
87. Line W.F., Kwong A., Weetal H.H.: Biochim. Biophys. Acta 242, 194 (1971).
88. Nishikawa A.H., Bailon P.: Arch. Biochem. Biophys. 168, 576 (1975).
89. Borisova V.N., Lomako O.V., Molina L.I., Nakhapetyan L.A.: Prikl. Biokhim. Mikrobiol. 15 (5), 744 (1979).
90. Balaskova O.B., Zubakova L.B., Korshak V.V., Nikiforova L. Ya.: Deposited Doc. VINITI 3742 (1978).
91. Franzmann G., Huelsmann H.L.: Ger. Offen. 2,821.890 (1979).

92. Gilpin R.K., Korpi J.A., Janicki L.A.: Anal. Chem. 47, 1498 (1975).
93. Regnier F.E., Noel R.: J. Chromatogr. Science 14, 316 (1976).
94. Jervis L.: Chrom. Synth. Biol. Polym. 2, 231 (1978).
95. Lowe C.R.: Int. J. Biochem. 8, 177 (1977).
96. Chang S.H., Gooding K.M., Regnier F.E.: J. Chromatogr. 125, 103 (1976).
97. Becker N., Unger K.K.: Chromatographia 12, 539 (1979).
98. Gruber K.A., Whitaker J.M., Morris M.: Anal. Biochem. 97, 176 (1979).
99. Barth H.D., Regnier F.E.: J. Chromatogr. 192, 275 (1980).
100. Schutyser J., et al.: Affinity Chromatography and Related Techniques, p. 143, Proc. of the 4th Int. Symp., Veldhoven (1981).
101. Robinson P.J.: Biochim. Biphys. Acta 249, 649 (1971).
102. Foucault A., Caude M., Oliveros L.: J. Chromatogr. 185, 345 (1979).
103. Saeed Talat, Sanara P., Vercela M.: Adv. Chromatogr. (Houston) 1979 (14), 669.
104. Matlin S.A., Tinker J.S.: J. High Resolut. Chromatogr. Commun. 2, 507 (1979).
105. Lin Chien K., Lee Ching S., Perrin J.H.: J. Pharm. Sci. 69, 95 (1980).
106. Parr W., Grohmann K.: Tetrahedron Lett. 28, 2633 (1971).
107. Roumeliotis R., Unger K.: J. Chromatogr. 185, 445 (1979).
108. Machleidt W.: Proc. Int. Conf. Solid-phase Methods in Protein Sequence Analysis 17 (1975).
109. Bridger J.: FEB Lett. 50, 159 (1975).
110. Page M.I., Jencks W.P.: J. Amer. Chem. Soc. 94, 8818 (1972).
111. Birr Chr. (Ed.): Methods in Peptide and Protein Sequence Analysis, Elsevier Amsterdam 1980.
112. Rony P.R.: J. Catalysis 14, 142 (1969).
113. Robinson K.K., Paulik F.E., Hershman A., Roth J.F.: J. Catalysis 15, 245 (1969).
114. Ward M.D., Schwartz J.: J. Mol. Catal. 11, 397 (1981).
115. Ward M.D., Schwartz J.: J. Amer. Chem. Soc. 103, 5253 (1981).
116. Niebergall H.: Makromol. Chem. 52, 218 (1962).
117. Czaková M., Čapka M.: J. Mol. Catal. 11, 313 (1981).
118. Marciniec B., Koznetka Z.W., Urbaniak W.: J. Mol. Catal. 12, 221 (1981).

119. Bartholin M., Conan J., Guyot A.: J. Mol. Catal. 2, 307 (1977).
120. Allum K.G., Hancock R.D., McKenzie S., Pitkethly R.C.: 5th Int. Congress on Catalysis, West Palm Beach, Florida 1972.
121. Allum K.G., Hancock R.D., Howell I.V., Lester T.E. et al.: J. Catal. 43, 331 (1976).
122. Allum K.G.: J. Organometal. Chem. 107, 393 (1976).
123. Bartholin M., Graillat C., Guyot A.: J. Mol. Catal. 10, 99 (1980).
124. Čapka M.: Synt. React. Inorg. Met.-Org. Chem. 7, 347 (1977).
125. Kozák Z., Čapka M.: Collect. Czech. Chem. Commun. 44, 2624 (1979).
126. Kavan V., Čapka M.: Collect. Czech. Chem. Commun. 45, 2100 (1980).
127. Michalska Z.M., Čapka M., Stoch J.: J. Mol. Catal. 11, 323 (1981).
128. Kolb I., Černý M., Hetflejš J.: React. Kinet. Catal. Lett. 7, 199 (1977).
129. Whitehurst D.D.: Chem. Technol. 1980, 44.
130. Yermakov Yu.I., Kuznetsov B.N., Zakharov V.A.: Catalysis by Supported Complexes, Elsevier, Amsterdam-New York (1981).

2

INTRAMOLECULAR INTERACTION IN THE CHEMICAL BEHAVIOR OF CARBON-FUNCTIONAL ORGANOSILICON COMPOUNDS

Josef Pola

Institute of Chemical Process Fundamentals
Czechoslovak Academy of Sciences
Prague, Czechoslovakia

2.1 INTRODUCTION

In many respects the unusual behavior of carbon-functional organosilicon compounds, i.e., compounds in which silicon is bonded to an sp^3 hybridized carbon of an organic framework containing a functional group,

$$Si\text{---}(C)_n\text{---functional group}$$

[functional group = a group bonded to the $(C)_n$ framework by an atom other than a formally neutral sp^3 carbon] is associated with a greater atomic volume, a lower electronegativity, and more polarizable bonds of silicon compared to carbon (Chapter 4, Section 4.4). The greater atomic volume of silicon makes steric shielding less significant; in silacyclic compounds the greater atomic volume adds conformational effects that change torsional angles and transannular repulsions between non-bonded substituents of the ring. The lower electronegativity of silicon in trialkylsilyl substituted compounds creates a negative charge on the adjacent carbon, and the inductive mechanism provides the following charge distribution in a silyl-alkyl framework

$$R_3Si\text{---}\!\!\rightarrow\overset{\delta-}{C}\text{---}\!\!\rightarrow\overset{\delta\delta-}{C}\text{---}\!\!\rightarrow\overset{\delta\delta\delta-}{C}$$

The opposite charge distribution is to be expected in this system when alkyl groups on silicon are replaced by more electronegative substituents.

$$\text{Cl}_3\text{Si} \leftarrow\text{---}\overset{\delta+}{\text{C}}\leftarrow\text{---}\overset{\delta\delta+}{\text{C}}\leftarrow\text{---}\overset{\delta\delta\delta+}{\text{C}}$$

This inductive mode of electron displacement is often accompanied by other types of intramolecular interactions. Thus, in $R_3\text{Si-C-C}^{\delta\pm}$ systems a hyperconjugative mode of electron displacement becomes important (Chapter 4, section 3). The silicon atom appears able to stabilize reaction intermediates either by donating electrons to an electron deficient β-carbon,

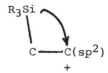

or by withdrawing electrons from this atom when it is electron rich.

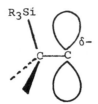

The importance of this conjugative mechanism increases with the electronic demands of the reaction center.

Another mode of intramolecular interaction occurring in carbon-functional organosilicon compounds (carbon-functional silanes) is through-space interaction between silicon and a functional group (Chapter 4, section 4.5) or between silicon and a reaction center created in, e.g., an intermediate species. Intramoleuclar interactions take place in compounds in which the functional group can approach the silicon easily (n > 2), and in compounds (n = 1) where such an interaction is impeded by some distortion of the SiC (functional group) bond angle.

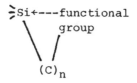

INTRAMOLECULAR INTERACTION

In some instances this interaction becomes important only during reactions; in others it operates even in the ground state. These effects all produce not only a different reactivity, but the reaction for carbon-functional silanes is also often different from that of carbon analogues.

In this chapter an attempt is made to describe the chemical behavior of carbon-functional organosilicon compounds as a result of intramolecular interactions. Electronic (inductive, conjugative, anchimeric assistance) and steric effects on the reactivity of carbon-functional silanes are discussed in Section 2 for those reactions in which carbon-functional silanes and their carbon analogues follow the same pathway. Reactions of carbon-functional silanes that differ from carbon analogues in their pathways are presented in Section 3. These reactions are dominated by a through-space interaction, which occurs between the silicon and a functional group, and which originates from a coordinative interaction between the silicon and a functional group center that has nucleophilic character. These reactions, unknown in carbon analogues, involve intramolecular reorganization in which products with a functional group directly bonded to silicon are formed. The reorganization is facilitated by thermodynamics; the bonds formed are stronger than those that are broken. In some of these reactions a migration of a silyl group from a carbon to a functional group occurs; in others, formation of a bond between the silicon and the functional group is accompanied by elimination of the ≡Si-functional group molecule.

This chapter omits reactions of carbon-functional silanes that differ in pathway from their carbon analogues due to a special effect of the silicon itself, such as its greater tendency to act as an electrofug [1-3], where an electrofug is defined as a leaving group that is electron-deficient

$$Nu:^- + R_3Si-C-C-Y \longrightarrow NuSiR_3 + {>}C{=}C{<} + Y^-$$

or as an electrophilic molecular center [4]. (The $RCH\overset{O}{-}CH_2$ molecule undergoes hydride attack at the sterically more accessible carbon atom and yields $RCH(OH)Me$.)

$$R_3Si-CH-CH_2 \xrightarrow{LiAlH_4} \left[R_3Si-CH-CH_2 \right] \xrightarrow{H_2O} R_3SiCH_2CH_2OH$$

(with epoxide oxygen, hydride from AlH₃ attacking, and electrophile E)

Also omitted are reactions involving a transition state formed by two intermolecular interactions in which both the silicon and the functional group take part [5],

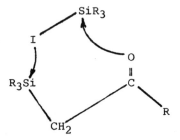

or, alternatively, the interaction occurs between the silicon and the carbon bearing the functional group [6].

$$Me_3Si \cdots CH_2 - Ph$$
$$\vdots \vdots$$
$$Cl \cdots AlCl_2$$

2.2 SUBSTITUENT EFFECTS OF SILYL GROUPS IN ORGANIC REACTIONS*

The substituent effects of the silyl group include the electronic effect, (i.e., the inductive effect, the conjugative effect, and anchimeric assistance), and it is discussed in connection with the steric effect when necessary. Each presentation and discussion of pertinent data on the substituent effects of silyl groups in organic reactions is preceeded by a brief outline of the mechanism, or at least the rate-determining step, of the appropriate reaction.

2.2.1 Reactions of Organometallic Compounds

The comparative reactivity of $R_nX_{3-n}Si(CH_2)_m$-metal compounds has not been studied thus far. The α- and β-silylated compounds show an increased stability that can be ascribed to cooperation of steric and electronic effects in the former, and to electronic effects only in the latter.

<u>Increased Stability of α-Silylated Compounds</u>. There are numerous examples of increased stability of organometallic compounds bearing trimethylsilyl group(s) on the α-carbon.

The higher thermal stability [7] of tetrakis-trimethylsilylmethyl transition metal alkyls, $(Me_3SiCH_2)_4M$, can be attributed [8] (1) to the absence of β-hydrogen atoms in R, which precludes the olefin-elimination decomposition pathway, and (2) to the large steric hindrance of the coordination site by bulky R groups, thereby making inter- or intra-molecular reactions into high activation energy processes.

A remarkable thermal stability and a reluctance to undergo reactions at the metal center are properties of tris(trimethylsilyl)methyl-metal derivatives. The trisyl [$(Me_3Si)_3C$] group causes a very large steric hindrance to nucleophilic displacements at the silicon to which it is

---
* For a theoretical treatment of this phenomenon see Chapter 4.

attached, and most of the common displacements at
silicon do not occur. Thus, the $(Me_3Si)_3C-SiCl_3$ compound
[9] does not react with boiling alcoholic silver nitrate,
and the $(Me_3Si)_3C-SiL_2X$ (L = Me or Ph) compounds give [10]
no substitution products on treatment with boiling
MeONa-MeOH.

Similarly, the ability of $[(Me_3Si)_3C]_2Zn$ and
$[(Me_3Si)_3C]_2Cd$ to withstand boiling in aqueous tetrahydrofuran is the most unusual feature of the chemistry of
these compounds [11].

The remarkable ease of formation and the stability of
$(Me_3Si)_3C-Li$ in tetrahydrofuran may be associated [12]
with stabilization of the $(Me_3Si)_3C^-$ carbanion by delocalization of the lone pair electrons due to the highly
polarizable bonds of the silicon atom. This explanation
is consistent with the high kinetic acidity of
$(Me_3Si)_3C-^3H$, which undergoes [13] detritiaton about 5-7
times as rapidly as Ph_3C-^3H.

Increased Stability of β-Silylated Compounds. Unexpectedly high stability can also be attributed to organomagnesium compounds with a triphenylsilyl group in the
β-position. Treatment of 2-(bromoethyl)triphenylsilane
with phenylmagnesium bromide in tetrahydrofuran leads to
extensive halogen-metal exchange (Eq. 1), which is not a
common process with the organomagnesium reagent

$$Ph_3SiCH_2CH_2Br + PhMgBr \rightleftharpoons Ph_3SiCH_2CH_2MgBr + PhBr \quad (1)$$

and has been reported only for alkyl halides with electronegative substituents stabilizing the carbanionic state
[14]. The $Ph_3SiCH_2CH_2^-$ carbanion appears to be more
stable than the phenyl carbanion, which can be explained
either by a conjugative mechanism or by a three-center
intramolecular interaction between the carbanionic center
and the silicon.

2.2.2 Reactions of Alkenes

Addition reactions of alkenes that are presented and
discussed here are divided into (1) heterolytic additions
that proceed by an ionic mechanism with an electrophilic
attack of a reagent on the double bond as the first step,

and (2) a radical mechanism involving homolytic addition.

<u>Heterolytic Additions</u>. Heterolytic additions are facilitated in their rate-determining electrophilic attack of the reagent on the double bond by the electron-donating effect of trialkylsilyl groups.

Addition reactions of hydrogen halides, bromine, chlorine, or thiocyanogen to the double bond of $Me_3Si(CH_2)_mCH=CH_2$ compounds are made easier [15,16] with a decrease in m, the number of intervening methylene groups. The activating, electron-releasing effect of the Me_3SiCH_2 group is noticeably higher than that of the $Me_3Si(CH_2)_2$ group, and the effect of the $Me_3Si(CH_2)_3$ group is comparable with that of groups with higher m.

The reactivity of allylsilanes, $Cl_nR_{3-n}SiCH_2CH=CH_2$, is controlled by the character of the silyl group, which also determines the direction of the polar reagent addition. An increasing number of chlorines retards [17] the addition of PCl_5. In the reaction shown in (Eq. 2), the acting electrophilic species is presumed to be the ionic pair $PCl_4^+ \ldots PCl_6^-$.

$$Cl_nMe_{3-n}SiCH_2CH=CH_2 + PCl_5 \xrightarrow{SO_2} Cl_nMe_{3-n}SiCH_2\underset{\underset{Cl}{|}}{CH}CH_2POCl_2 \quad (2)$$

The same retarding effect is exerted in the addition of hydrogen chloride, chlorine, and bromine. These reagents, which react very rapidly with $Me_3SiCH_2CH=CH_2$ even at low temperatures, react very slowly with $Cl_3SiCH_2CH=CH_2$. Addition of HCl to the latter compound does not take place at all.

The course of the addition of hydrogen halides to alkenes is controlled by the character of the $Cl_nMe_{3-n}SiCH_2$ group, as is shown in Table I.

With n = 0 - 2 (and for HX = HI with n = 3) the solvated proton of HX attacks the terminal carbon atom of the allylic chain (Markovnikov's rule observed), which results in the formation of $Cl_nMe_{3-n}SiCH_2CH(X)Me$. The central carbon of the allylic system is attacked by the proton when n = 3 and HX = HBr. Surprisingly, the addi-

Table I. Addition of hydrogen halides to $Cl_nMe_{3-n}SiCH_2CH=CH_2$ compounds [18-20]

n	HX	Admixture	
		–	Benzoyl peroxide
0	HCl	M[a]	M
	HBr	M	M
	HI	M	M
1	HBr	M	M
	HI	M	–
2	HBr	M	anti M
	HI	M	–
3	HCl	no reaction	–
	HBr	slow reaction anti-M	anti M
	HI	M	–

[a] Markovnikov's rule observed.

tion of HBr in the presence of benzoyl peroxide when n = 0 or 1 leads to the same products as heterolytic addition. Under these conditions the $Cl_nMe_{3-n}Si(CH_2)_3X$ compounds (anti-Markovnikov addition) are formed only for n = 2 and 3.

This orientation of the electrophile and the retarding effect of chlorine atoms on the reaction rate can be explained by the fact that heterolytic addition (n = 0,1) proceeds so rapidly that the free-radical reaction cannot compete with it. The decreasing electron-releasing effect of $Cl_nMe_{3-n}Si$ groups with increasing n weakens the nucleophilic ability of the alkene, which may lead not only to a slower heterolytic reaction, and in the presence of a radical initiator to a certain facilitation of the free-radical reaction, but it may also favor the heterolytic addition as depicted below.

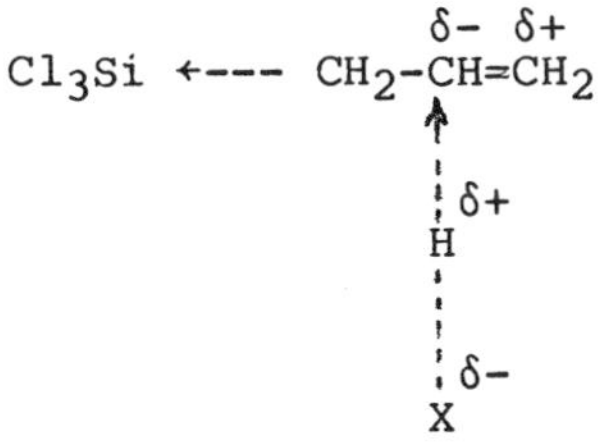

$$Cl_3Si \longleftarrow CH_2\overset{\delta-}{-}\overset{\delta+}{CH}=CH_2$$
$$\underset{X}{\overset{H}{|}}\,\delta+$$
$$\delta-$$

Free-Radical Additions. Reactions of these formally neutral species are understood to be influenced considerably by a polar substituent in a molecule with a double bond. The addition of radicals such as ·CCl$_3$ and singlet carbenes such as :CCl$_2$ and :CCO$_2$Et to alkenes are rationalized as additions of electrophilic reagents, and the reactivities of alkenes toward these species therefore decrease with their decreasing nucleophilic character. The reactions of silyl substituted alkenes are influenced by the nature of the transfer reagent.

Reactions With Less Reactive Transfer Agents. Reactions of $Cl_nR_{3-n}Si(CH_2)_mCH=CH_2$ with less reactive transfer agents such as CCl$_4$, CHCl$_3$, HCO$_2$H, RCHO or Cl$_3$SiH are slower [21] for n = 1 than for n = 2, the reactivity of the double bond for n = 1 being increased [22] with a pronounced electronegative nature of the substituents attached to silicon. The latter effect is confirmed by the yields of the Cl$_3$SiH addition to $R_nX_{3-n}SiCH_2CH=CH_2$ compounds in the presence of benzoyl peroxide:

$R_nX_{3-n}Si$:	Me$_3$Si	Me$_2$PhSi	Ph$_3$Si	Cl$_3$Si
Yield, %	2.5	22	29	62

The low reactivity of allyltrialkylsilanes in these reactions can be explained by the formation in the initial addition step of a highly stabilized radical (Eq. 3)

$$Me_3SiCH_2CH=CH_2 + \cdot CCl_3 \longrightarrow Me_3SiCH_2\text{-}\overset{\cdot}{C}H\text{-}CH_2CCl_3 \quad (3)$$

that is incapable of chain propagation, whereas with electronegative substituents on silicon the intermediate radical is relatively reactive.

Reactions With Highly Reactive Transfer Agents. Allylsilanes react rapidly with highly reactive transfer agents such as $BrCCl_3$. The energy barrier for chain propagation is low, and the reactivity is dominated by the ease of radical attack on the alkene. The addition of the $Cl_3C\cdot$ radical to alkenes [23] proceeds faster [24] in $Me_3Si(CH_2)_mCH=CH_2$ compounds than in $R-CH=CH_2$ compounds when m = 1, but when m increases, the rates of the organosilicon compounds approach those of the all-carbon compounds. The relatively high reactivity of $Me_3SiCH_2CH=CH_2$ cannot be explained only in terms of the +I effect of the Me_3SiCH_2 group, since $k_{rel}(Me_3SiCH_2CH=CH_2):k_{rel}(Me_3Si(CH_2)_2CH=CH_2)$ estimated [23] on this presumption is only 1:2. The much greater [24] rate enhancement for $Me_3SiCH_2CH=CH_2$ (Table II, reaction 2) apparently arises from an extra resonance effect in the intermediate free radical. The lower value of the ratio observed [25] for the same addition (reaction 2) under similar conditions is striking. It may be that there is insufficient development of radical character in the transition state for the influence of the resonance effect to be observable. The lower ratio was obtained

Table II. Relative rates of addition of radicals and carbenes to $Me_3Si(CH_2)_mCH=CH_2$ compounds

Reaction	Attacking species	k_{rel}				Ref.
		m=1	m=2	m=3	m=4	
1	$\cdot CCl_3$	1.00	0.19	0.13	0.11	24
2	$\cdot CCl_3$	1.00	0.66	0.55	0.48	25
3	$:CCl_2$	1.00	0.24	–	–	26
4	$:CHCO_2Et$	1.00	0.48	–	–	27
5	$:CPh_2$	1.00	0.97	–	–	28

when chlorobenzene was used as a solvent. Its interaction with the $Cl_3C\cdot$ radical thus remains a possible explanation of the observed differences. The highly activating, i.e., electron-releasing effect of the Me_3SiCH_2 group is also observed upon addition of the electrophilic dichloro- and carbethoxycarbene reagents, each of which reacts in its singlet state.

Diphenylcarbene, however, is a nucleophilic species [29] in its triplet (biradical) state. Its addition to alkenes substituted with electron-releasing $Me_3Si(CH_2)_m$ groups is therefore slow, contrary to olefins with a Me_3Si group bonded to the sp^2 carbon.

Table III shows how olefin reactivity toward the $Cl_3C\cdot$ radical is influenced by different silyl substituents when the number of CH_2 groups interposed between the silicon and the double bond increases. The reactivity of trimethylsilyl compounds decreases when m increases, which is consistent with a decreasing electron-releasing ability of $Me_3Si(CH_2)_m$ groups with increasing m.

In triphenylsilyl compounds, it was suggested that the $Cl_3C\cdot$ radical complexes with the phenyl ring. The reactivity is then assumed to be influenced not only by the polar effect of the Ph_3Si group, but also by the relative tendencies of the side chains to adopt conformations

Table III. Relative rates of addition of the $Cl_3C\cdot$ radical to alkenylsilanes [23]

Olefin	k_{rel}		
	m=1	m=2	m=3
$Me_3Si(CH_2)_mCH=CH_2$	1	0.66	0.55
$Ph_3Si(CH_2)_mCH=CH_2$	0.60	0.18	0.31
$Cl_3Si(CH_2)_mCH=CH_2$	0.31	0.32	0.33

bearing the double bond in the vicinity of the complexed $Cl_3C\cdot$ radical.

The low reactivity of trichlorosilyl compounds is constant for all values of m. This is not what one would expect, since the same decrease in the nucleophilic ability of an alkene by the -I effect of the Cl_3SiCH_2 group when m = 1, 2, and 3 is hardly understandable. It thus appears that other factors must be involved.

2.2.3 Reactions of Aromatic Compounds

<u>Electrophilic Substitutions</u>. These reactions proceed via a transition state that closely resembles a σ-complex, the formation of which is usually rate-determining. This mechanism, sometimes called the Wheland intermediate mechanism (Eq. 4), relates to all subsequently discussed reactions in which an electron-releasing substituent L will tend to enhance the reaction rate and to direct reaction to the ortho- and para-positions, while an electron-withdrawing substituent L will decrease the rate and is meta-directing.

<u>Detritiaton</u>. The effect of silicon (silyl substituents) on the reactivity of the aromatic ring in carbon-

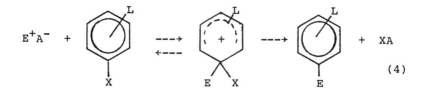

(4)

X = H, Me₃Si

functional silanes will be shown first in the most typical reaction for the investigation of aromatic reactivity; i.e., detritiation by trifluoroacetic acid. The data in Table IV show that $Me_3Si(CH_2)_m$ groups stabilize the Wheland intermediate and are rate-accelerating. This effect is considerably greater with the Me_3SiCH_2 group and is always more pronounced with substituents in the para position. The values of the log k_{rel}ortho/log k_{rel}para ratios for individual $Me_3Si(CH_2)_m$ groups correspond for all values of m to those shown empirically to apply to a wide range of monosubstituted benzenes [31].

Table IV. Rates of detritiation [30] of $L-C_6H_4-^3H$
by trifluoroacetic acid at 70°C.

No.	L	k_{rel}[a]	No.	L	k_{rel}[a]
1	p-Me	450	6	o-Me	220
2	p-Me$_3$SiCH$_2$	82000	7	o-Me$_3$SiCH$_2$	9300
3	p-Me$_3$Si(CH$_2$)$_2$	810	8	o-Me$_3$Si(CH$_2$)$_2$	450
4	p-Me$_3$Si(CH$_2$)$_3$	580	9	o-Me$_3$Si(CH$_2$)$_3$	270
5	p-Me$_3$Si(CH$_2$)$_4$	690		o-Me$_3$Si(CH$_2$)$_4$	270

[a] Relative to rates of [^{3}H] benzene.

Protodesilylation. Activating effects similar to those of Me$_3$Si(CH$_2$)$_m$ groups in electrophilic aromatic substitutions operate in the cleavage of Me$_3$Si(CH$_2$)$_m$C$_6$H$_4$SiMe$_3$ compounds by aqueous methanolic perchloric acid. The data given in Table V on this protodesilylation reaction reveal a very large activating effect of the p-Me$_3$SiCH$_2$ group that is markedly (about 10 times) greater than that of the p-Me and other p-Me$_3$Si(CH$_2$)$_m$ substituents. The m-Me$_3$SiCH$_2$ group activates the benzene ring meta position about 2.5 times as effectively as the m-Me group, and the activating effects of the other m-Me$_3$Si(CH$_2$)$_m$ groups are similar to each other and about 1.5 times as great as that of the m-Me group. A different situation exists with o-Me$_3$Si(CH$_2$)$_m$ substituents; the benzene ring position is activated (relative to o-Me) by the o-Me$_3$SiCH$_2$ group, but weakly deactivated by Me$_3$Si(CH$_2$)$_3$ and Me$_3$Si(CH$_2$)$_4$ groups. A comparison of the reactivity data for silicon and carbon analogues of the Me$_3$CCH$_2$C$_6$H$_4$SiMe$_3$ compounds, with data obtained under very similar conditions, reveals that the Me$_3$SiCH$_2$ substituent accelerates protodesilylation much more from the para position than does the Me$_3$CCH$_2$ substituent. The rates of protodesilylation with (Me$_3$Si)$_n$CH$_{3-n}$C$_6$H$_4$SiMe$_3$ compounds increase [34] in the order n = 0 < 3 < 1 < 2 for both the m- and p-compounds, a result that is not consistent with the attribution of the effects of the (Me$_3$Si)$_n$CH$_{3-n}$ group entirely to inductive electron release by the Me$_3$Si group.

Table V. Rates of protodesilylation [32-34] of L-C$_6$H$_4$-SiMe$_3$ by aqueous methanolic perchloric acid at 50°C

L	k_{rel}[a]	L	k_{rel}[a]	L	k_{rel}[a]
o-Me	17	m-Me	2.5	p-Me	21
o-Me$_3$SiCH$_2$	31	m-Me$_3$SiCH$_2$	6.6	p-Me$_3$SiCH$_2$	270
o-Me$_3$Si(CH$_2$)$_2$	17	m-Me$_3$Si(CH$_2$)$_2$	3.6	p-Me$_3$Si(CH$_2$)$_2$	28
o-Me$_3$Si(CH$_2$)$_3$	12	m-Me$_3$Si(CH$_2$)$_3$	3.8	p-Me$_3$Si(CH$_2$)$_3$	22
o-Me$_3$Si(CH$_2$)$_4$	13	m-Me$_3$Si(CH$_2$)$_4$	3.6	p-Me$_3$Si(CH$_2$)$_4$	24
-		m-Me$_3$CCH$_2$	4.1	p-Me$_3$CCH$_2$	21
-		m-(Me$_3$Si)$_2$CH	8.4	p-(Me$_3$Si)$_2$CH	670
		m-(Me$_3$Si)$_3$C	3.4	p-(Me$_3$Si)$_3$C	200

[a] Relative to rate of Ph-SiMe$_3$.

Nitration. The nitration [35] of Me$_3$Si(CH$_2$)$_m$Ph compounds in acetic anhydride is an interesting variation of the effects of trimethylsilylalkyl groups with a different number of methylene units separating the silicon atom from the benzene ring. The results (Table VI) show the high reactivity of the ortho (compared to the para) position in the n = 1 compound, the sharp fall in reactivity from the compound with n = 1 to that with n = 2, and a small rise in reactivity from the compound with n = 3 to that with n = 4.

Bromination. Bromination of L-Ph compounds with L = Et$_n$Cl$_{3-n}$Si(CH$_2$)$_2$ (n = 0, 1, and 3) and with L = Me$_n$Cl$_{3-n}$SiCH$_2$ (n = 0-2) in the presence of iron gives [36] solely p-substituted products. With L = Me$_3$SiCH$_2$ the meta and para isomers, however, are formed in equimolar amounts, which, in view of the activating effect of the Me$_3$SiCH$_2$ group, is a striking result. Table VII shows that the activating effect of Me$_3$Si(CH$_2$)$_m$ groups follows the order m = 1 > m = 2, as in previous electrophilic substitutions. However, for trichlorosilyl groups the reverse order is valid.

Table VI. Rates[a], k_{rel}, and partial rate factors [32], f, of nitration [35] of $Me_3Si(CH_2)_mPh$ compounds in acetic anhydride at -10°C

m	k_{rel}	f_{ortho}	f_{para}
1	77.2	197	70
2	8.9	12	29
3	10.6	16	32
4	19.0	37	40

[a] Relative to rate of benzene.

Acetylation. The same order of the activating effect of $Me_3Si(CH_2)_m$ groups is reflected by the rates of acetylation [37] of L-Ph compounds by acetic anhydride catalyzed by aluminum chloride:

L:	H	Me_3SiCH_2	$Me_3Si(CH_2)_2$
k_{rel}:	1	24	16

Discussion. The reactivity data of trimethylsilyl-alkyl substituted benzenes, $Me_3Si(CH_2)_mPh$, in electrophilic substitution show that the greatest activating effect is exerted by the Me_3SiCH_2 group. In detritiation and protodesilylation this group stabilizes the p-position of a σ-complex much more effectively than the o-position. The reverse stabilizing effect for these positions in nitrations can be ascribed [31] to an interaction of the silicon atom with an incipient nitrate ion from dinitrogen pentoxide in the transition state. The activating effects of p- and m-$Me_3Si(CH_2)_m$ groups in detritiation and protodesilylation are very similar, which is reflected [30] by a linear correlation between the logs k_{rel} of p- and m-trimethylsilylalkyl substituted compounds for both reactions (Fig. 1). In this plot all points lie close to a straight line except for the o-Me_3SiCH_2 substituent. The anomalously low

Table VII. Rates of bromination [36] of L-Ph compounds

L	k_{rel}[a]	Isomer Formed
Me_3SiCH_2	130	meta + para
$Me_3Si(CH_2)_2$	11	?
$Cl_3Si(CH_2)_2$	5.8	para
Cl_3SiCH_2	4.7	para

[a] Relative to rate of benzene.

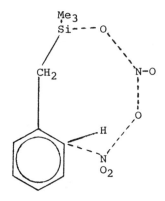

reactivity of the o-Me_3SiCH_2 compound in protodesilylation is likely associated with some steric hindrance to ortho substitution in this reaction. The great activating effect of the Me_3SiCH_2 group relates to its very high electron-releasing ability. The mechanism of the electron release is an electron delocalization [34,38-40] from the Si-C bond that is significant in the ground state of molecules [30] and can even be more pronounced in the transition states of electrophilic aromatic substitutions by the electron demand of the reaction center.

Dissociation of Benzoic Acids. The apparent dissociation constants of silylalkyl substituted acids such as L-C_6H_4-CO_2H (Table VIII) reveal [41,42] that the acidity of

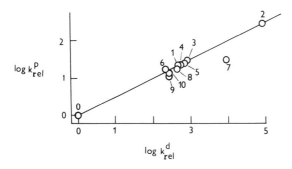

Fig. 1. Plot [30] of the logarithms of the relative rates (k_{rel}) of protodesilylation (p) against that of detritiation (d) for $L-C_6H_4-K$ compounds (K = $SiMe_3$ or ^3H). For numbering denoting L (0 = H) see Table IV.

benzoic acid is decreased upon substitution by a trialkylsilyl group. The operation of the large electron-releasing effect of the Me_3SiCH_2 group falls off in the order o > m > p, and of all of the silylalkyl groups studied, the decrease is the greatest for the p-$(Me_3Si)_2CH$ group. The Hammett substituent constants (σ) derived from the acidities of p-$(Me_3Si)_nCH_{3-n}$ groups are given in Table IX.

Table VIII. Apparent dissociation constants of $L-C_6H_4-CO_2H$ in 50% aqueous ethanol [41,42]

L	pK_a	L	pK_a
H	5.70	p-Me	5.88
o-Me	5.70	p-Me_3SiCH_2	6.08
o-Me_3SiCH_2	6.34	p-$(Me_3Si)_2CH$	6.14
m-Me	5.77	p-$(Me_3Si)_3C$	6.05
m-Me_3SiCH_2	6.00	p-OMe	6.02

Solvolysis of Cumyl Chlorides. The solvolysis of silylmethyl substituted cumyl chlorides $L-C_6H_4-CMe_2Cl$ in aqueous acetone [42] proceeds by a dissociative heterolysis of the C-Cl bond (S_N1 mechanism), and its rate is controlled by its first step, which is the formation of a carbonium ion (Eq. 5).

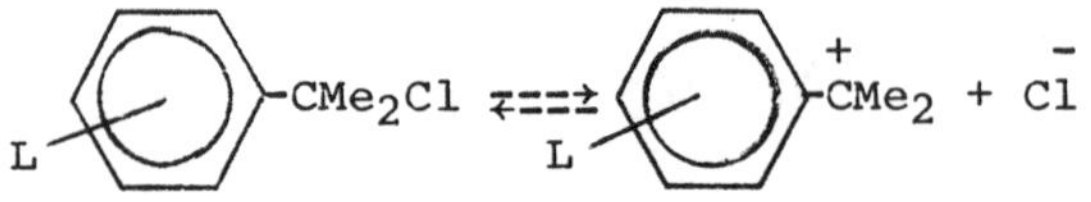

$$L-C_6H_4-CMe_2Cl \rightleftharpoons L-C_6H_4-\overset{+}{C}Me_2 + Cl^- \quad (5)$$

The rate constants of solvolysis of $L-C_6H_4-CMe_2Cl$ compounds given in Table IX thus reflect the ease of generation of carbonium ions from these chlorides. The rate constant of this reaction is best correlated with the substituent parameter σ^+, and hence σ^+ constants of $(Me_3Si)_nH_{3-n}C$ groups could be calculated from the reactivities of silylmethyl substituted cumyl chlorides. The $(\sigma^+-\sigma)$ values for these groups lie between those of the p-Me and p-OMe groups and indicate [42] that trimethylsilylmethyl groups have substantial electron-releasing conjugative effects that originate in hyperconjugation of the Me_3Si-C bond.

Table IX. Rate constants of solvolysis of $p-L-C_6H_4CMe_2Cl$ compounds in 90% aqueous acetone, and substituent parameters of L groups [42]

L	k_{rel}	σ	σ^+	$(\sigma^+-\sigma)$
Me	25	-0.17	-0.31	-0.14
Me_3SiCH_2	279	-0.29	-0.54	-0.25
$(Me_3Si)_2CH$	682	-0.33	-0.62	-0.29
$(Me_3Si)_3C$	222	-0.27	-0.52	-0.25
MeO	3360	-0.27	-0.78	-0.51

[a] Relative to rate of $PhMe_2CCl$.

2.2.4 Reactions of Compounds With Keto, Carboxyl, and Carbalkoxyl Groups

Compounds with the structure silyl-$(C)_n$-$C(O)L$, where L = R, OH, and OR, are discussed in this section. Compounds having a silyl group attached to a carbon within an R group are presented in Section 2.2.6.

<u>Dissociation of Cyanohydrins of Ketones.</u> The reactivities of carbon-functional silanes, which are influenced by the silicon conformational effect, are apparent from a comparison of cyanohydrin dissociation constants for various cyclohexanones (Table X). The enlargement of a six-membered ring by the substitution of a silicon atom for carbon will increase the hydrogen-hydrogen repulsions. Internal strain, however, is decreased in the cyanohydrin dissociation (Eq. 6).

Table X. Cyanohydrin dissociation constants [43] at 25°C

Ketone	$K_D \cdot 10^4$ mole/L
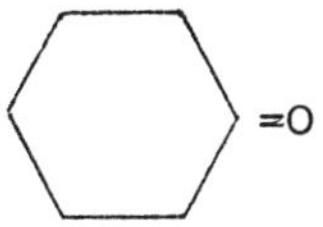 cyclohexanone	8.3
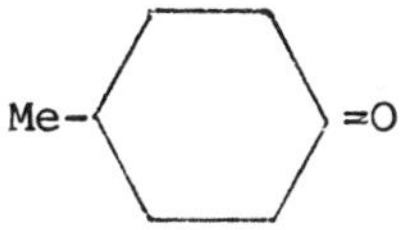 4-methylcyclohexanone	4.9
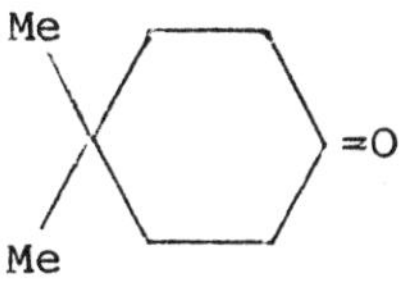 4,4-dimethylcyclohexanone	20.8
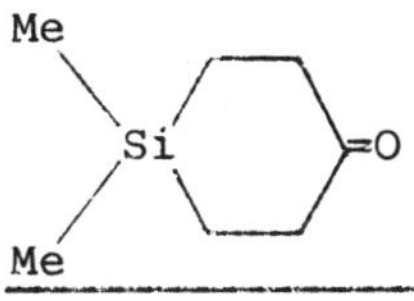 4,4-dimethyl-4-silacyclohexanone	94.7

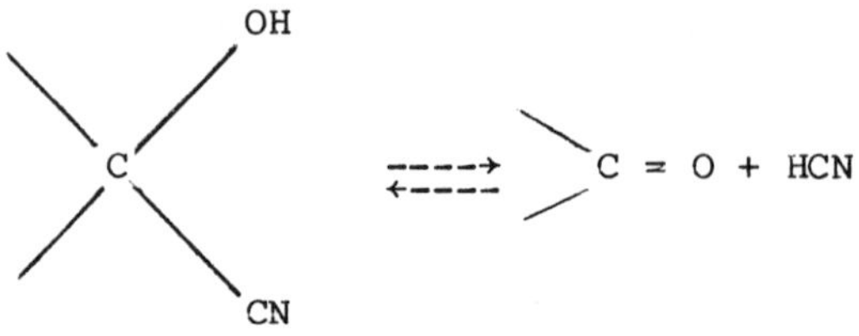

$$\diagdown_{\diagup}\!\!\!C\!\!\diagup^{OH}_{\diagdown CN} \quad \xrightleftharpoons[]{} \quad \diagdown_{\diagup}C = O + HCN \qquad (6)$$

and is a reason why the dissociation to silacyclohexanone will be favored over that of cyclohexanones [43].

Dissociation of Carboxylic Acids. Silyl substituted carboxylic acids are weaker [44] than the structurally similar carboxylic acids (Table XI). The acid-weakening effect of organosilyl substituents shows these substituents to operate via a +I effect that diminishes in a saturated carbon chain.

Hydrolysis of Esters. The rate-determining step of alkaline hydrolysis of primary alkyl esters, which is the attack of a hydroxyl anion on the carbonyl carbon of the ester, is facilitated by an electron deficiency and hindered by a steric shielding of this center.

The base-catalyzed hydrolysis of triethylsilylalkyl acetates (Table XII) is retarded by electron-releasing

Table XI. Dissociation constants [44] for LCH_2CO_2H at 25°C

L	$K \cdot 10^5$
H	1.75
Me	1.32
Me_3C	1.00
Me_3Si	0.60
$PhMe_2Si$	0.54
Me_3SiCH_2	1.24

substituents, and the relatively high rate of hydrolysis of the substituted propionate, compared to that of the butyrate, is not consistent with the expectation that the +I effect of the Et_3Si group should be more apparent if it is in a position closer to the carbethoxy group. The high reactivity of the former compound was explained [45] by an interaction between the carbonyl oxygen and the silicon in the transition state of the reaction:

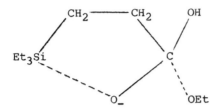

2.2.5 Reactions of Amines

The comparative reactivities of silylalkylamines in different reactions have not been studied except for the dissociation of conjugate acids of amines and the complexation of amines with acids and bases. This information, however, is useful in considering the reactivities of silylalkylamines in other reactions.

Acid-Base Equilibria. The aqueous and alcoholic base strengths of trialkylsilylalkylamines, $R_3Si(CH_2)_mNH_2$ and $R_3Si(CH_2)_mNR_2$, with m = 1 and 2, are appreciably higher than those of parent alkylamines (Table XIII). Thus, $Me_3SiCH_2NH_2$ is 5.7 times as strong as its carbon counterpart, neopentylamine, and almost twice as strong as n-propylamine [46]. The base strength of $R_nR_{3-n}^xSiCH_2NH_2$ compounds decreases with the increasing electron-withdrawing ability of the R^x group, but it is always higher than that of similar carbon compounds [50]. A similar increase in basicity is also observed for β-trialkylsilylethylamines. An examination of Table XIII reveals that introduction of higher alkyls into the R_3Si group of α-, β-, and γ-silylalkylamines results in a decrease of the amine basicity in all cases. The γ-trialkylsilylalkylamines possess a basicity comparable to their carbon analogues. The aqueous and alcoholic base strengths are consistent with the electron-releasing

Table XII. Rate constants for the alkaline hydrolysis of some esters (L mole^{-1} sec^{-1}) in aqueous ethanol [45]

Ester	$k \cdot 10^3$	Ester	$k \cdot 10^3$
MeCO$_2$Et	7.40	Et$_3$SiCH$_2$CO$_2$Et	-
MeCH$_2$CO$_2$Et	3.63	Et$_3$Si(CH$_2$)$_2$CO$_2$Et	2.43
Me(CH$_2$)$_2$CO$_2$Et	1.87	Et$_3$Si(CH$_2$)$_3$CO$_2$Et	1.51

ability of Me$_3$SiCH$_2$ and Me$_3$Si(CH$_2$)$_2$ groups; however, the somewhat lower basicity for α- vs. β-trialkylsilylalkylamines is staggering. It was assumed that the greater +I effect of the Me$_3$SiCH$_2$ group is compensated by its steric effect working against solvation of the protonated amine.

The latter effect is less significant, however, in controlling the complexation of amines with deuteriochloroform in CCl$_4$. Nevertheless, the basicity order shown below still holds and is valid in general for (EtO)$_n$Me$_{3-n}$Si(CH$_2$)$_m$NH$_2$ compounds (Table XIV).

$$\beta > \gamma, \delta > \alpha$$

The decrease in basicity when going from β- to α-silylalkylamines also holds [52] for silylalkylpiperidines L-NC$_5$H$_{10}$:

L:	H	n-Pr	Et$_3$SiCH$_2$	Et$_3$Si(CH$_2$)$_2$
$\Delta\nu_{C-D}$, cm^{-1}	65	70	60	80

The explanation was recently [53] provided by intrinsic solvent-free equilibrium constants for proton transfer reactions between amines and various bases determined by ion cyclotron resonance mass spectrometry (Table XV). These data confirm the electron-releasing properties of Me$_3$Si(CH$_2$)$_n$ (n = 1, 2) groups and show homologation to produce identical changes in both the carbon and the silicon systems. The lower basicity of Me$_3$SiCH$_2$NH$_2$ compared to its

INTRAMOLECULAR INTERACTION

Table XIII. Dissociation constants pK_a of conjugate acids of trialkylsilylalkylamines $L-NR_1R_2$ [46-49] at 25°C

L	NR_1R_2			
	NH_2[a]	$N(CH_2)_3CH_2$[b]	$N(CH_2)_4CH_2$[b]	$N(CH_2)_5CH_2$[b]
n-Pr	10.69	10.45[c]	10.32	10.80[c]
Me_3CCH_2	10.21	10.40	9.86	10.00
Me_3SiCH_2	10.96	10.77	10.62	11.02
$Me_3Si(CH_2)_2$	10.99	10.83	10.68	11.03
$Me_3Si(CH_2)_3$	10.75	10.79	10.33	10.79
Et_3SiCH_2	-	10.56	10.40	-
$Et_3Si(CH_2)_2$	-	10.74	10.52	-
$Et_3Si(CH_2)_3$	-	10.42	10.20	-

[a] In H_2O
[b] In MeOH
[c] L = Et

β-homolog, which is not consistent with the greater +I effect of the Me_3SiCH_2 group, appears to result from an intramolecular interaction between the silicon and nitrogen.

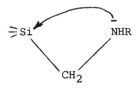

Such an interaction weakens the basicity by an electron transfer from nitrogen to silicon. It can also explain the relative acidities of silylmethylamines measured as the frequency shift of the N-H bond upon complexation of the amines with tetrahydrofuran [54]:

Table XIV. $CDCl_3$ frequency shifts[a], $\Delta\nu_{C-D}$ in cm^{-1}, due to interaction with silylalkylamines, $(EtO)_n Me_{3-n}Si(CH_2)_m NH_2$ [51]

m	n	$\Delta\nu$
1	0	36
1	1	38
1	2	35
1	3	42
2	0	46
2	1	–
2	2	45
2	3	46
3	0	42
3	1	43
3	2	41
3	3	39

[a] $\Delta\nu$ for $n\text{-BuNH}_2 = 43 cm^{-1}$.

Compound:	$MeOMe_2SiCH_2$-NRH	$MeOMe_2SiCH_2$-N(CO_2Me)H	R_2NH
$\Delta\nu_{N-H}$, cm^{-1}	~140	124	20

These data show secondary silylmethylamines, $R_n X_{3-n}SiCH_2N(R)H$, to be as acidic as secondary amines that have one strongly electron-accepting group directly attached to the nitrogen, and much more acidic than secondary alkylamines. The high acidity of silylmethylamines obviously reflects a high mobility of the N-H hydrogen, for which an intramolecular interaction as depicted on the previous page can be responsible.

Table XV. Gas-phase proton affinities of tertiary carbon and silicon amines [53]

Amine	Proton affinity,[a] J/mol	Amine	Proton affinity,[a] J/mol
$Me_3SiCH_2NMe_2$	950.9	$Me_3CCH_2NMe_2$	945.4
$Me_3Si(CH_2)_2NMe_2$	953.0	$Me_3C(CH_2)_2NMe_2$	946.3
$Me_3Si(CH_2)_3NMe_2$	953.0	-	

[a] Based upon proton affinity of NH_3 = 842 ± 8 J/mol.

2.2.6 Reactions of Alcohols, Their Esters, Acetals, and Ethers

Compounds with a silyl substituent attached to the carbon of the alkyl group bonded to oxygen are presented in this section.

The way in which the reactivity of carbon-functional $X_nR_{3-n}Si(CH_2)_mO-$ silanes (with the sp^3-hybridized oxygen as a part of functional group) is influenced by the silyl group will depend on the distance between these groups and the nature of the reagent. Silyl groups separated from the oxygen by more than two CH_2 units have a small but still apparent effect only when m = 3. The $Me_3Si(CH_2)_2$ group affects the oxygen by its weak +I effect. The oxygen experiences the polar effect of the Me_3SiCH_2 group and appears to differ during interaction of the ethers with electrophiles and with nucleophiles. In the former reactions, the Me_3SiCH_2 group exerts a +I effect; in the latter reactions the Me_3SiCH_2 group is weakly electron-accepting, for which an O→Si intramolecular interaction in the ground state can be responsible.

<u>Addition of Alcohols to a Carbonyl Bond</u>. Addition of alcohols, ROH, to phenyl isocyanate or to ketene includes the rate-determining nucleophilic attack of the ROH molecule on the carbonyl carbon.

The relative reaction rates of $Me_3Si(CH_2)_mOH$ compounds in reactions with both reagents sharply decrease [55] when the OH group is moved from the α- to the β-position but remain unaltered with further removal from the γ- to the δ-position (Table XVI). The Me_3SiCH_2 group apparently increases the electron density on the oxygen and thus makes it more receptive to nucleophilic attack. This activating effect is not apparent when this group is separated from the reaction center by more than one $-CH_2-$ unit.

Displacement of Alcohol Hydroxyl by Halogen. Although no comparative study of reactivities of PY_3 and $SOCl_2$ with silyl substituted alcohols, $R_3Si(CH_2)_mOH$, was performed, it is quite clear that the β-silyl group exerts a specific effect in the first step of this reaction, which is the formation of an ester, e.g., ROSOY.

The reaction of $Me_3SiCH_2CD_2OH$ with the above reagents yields [56] a mixture of the α- and β-deuterated products

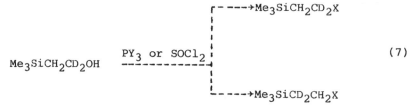

(7)

the presence of which was accounted for [56] by the rearrangement by the Me_3Si group in an ion pair.

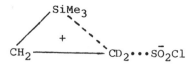

The occurrence of this rearrangement indicates that silicon interacts with the developing positive charge on the β-carbon atom and possibly provides anchimeric assistance.

Acid-Catalyzed Rearrangement of β-Heterosubstituted Alcohols. The acid-catalyzed dehydration of alcohols having PhS-, $Ph_2P(O)-$, or alkyl groups (generally a G group) in the β-position involves 1,2-migration of these groups. This migration is encouraged and the reaction

Table XVI. Relative reaction rates [55] of $Me_3Si(CH_2)_mOH$ with CH_2CO and PhNCO at 20°C

Reactant	n			
	1	2	3	4
CH_2CO	3.1	1.6	1.0	1.0
PhNCO	4.2	1.6	1.0	1.0

outcome controlled by the presence of the silyl group in the γ-position [57-59].

The acid-catalyzed rearrangement of β-(hydroxyalkyl)-phenyl sulfides leads [57] to phenylthio migration from a tertiary or secondary center to a primary one (Eq. 8).

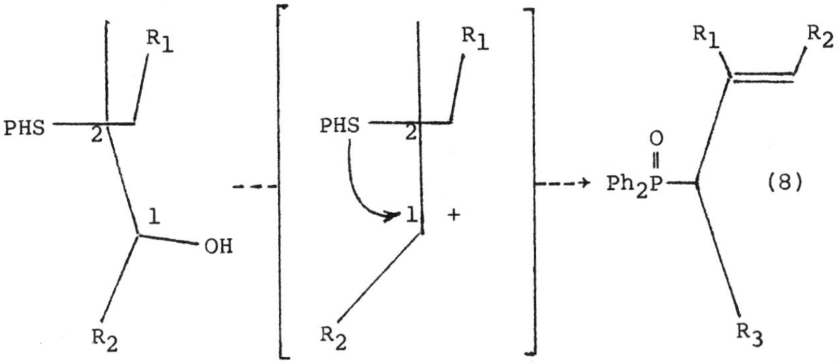

When assisted by silicon, migration between two secondary centers or even from a secondary to a tertiary center takes place.

The acid-catalyzed conversion of β-(hydroxyalkyl)-diphenylphosphine oxides into alkylphosphine oxides by diphenylphosphinyl migration shows [58] complete regioselectivity in favor of the more substituted olefin when the migration origin is unsymmetrical (Eq. 9).

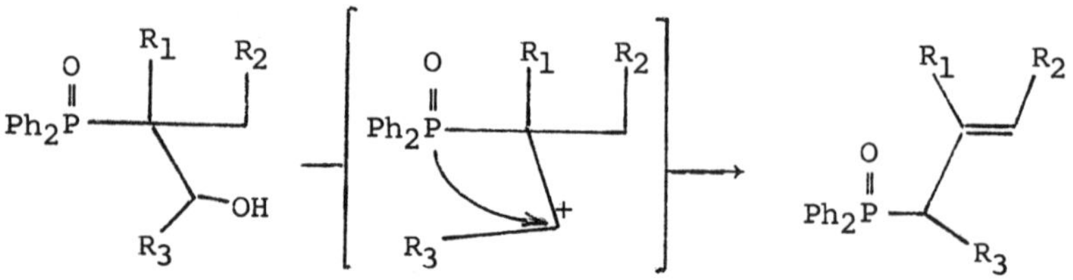

$$Ph_2\overset{O}{\overset{\|}{P}}\underset{R_3}{\overset{R_1\ R_2}{-C-C-}}OH \longrightarrow \left[Ph_2\overset{O}{\overset{\|}{P}}\underset{R_3}{\overset{R_1\ R_2}{-C-C-}}^+ \right] \longrightarrow Ph_2\overset{O}{\overset{\|}{P}}\underset{R_3}{-C} \overset{R_1\quad R_2}{\underset{}{=}} \quad (9)$$

The other less substituted olefin is formed exclusively in the presence of a γ-Me$_3$Si group that is lost in preference to a proton.

The reactions of alcohols having a silyl group in the γ-position will in general proceed markedly faster. The driving force of these reactions (Eq. 10)

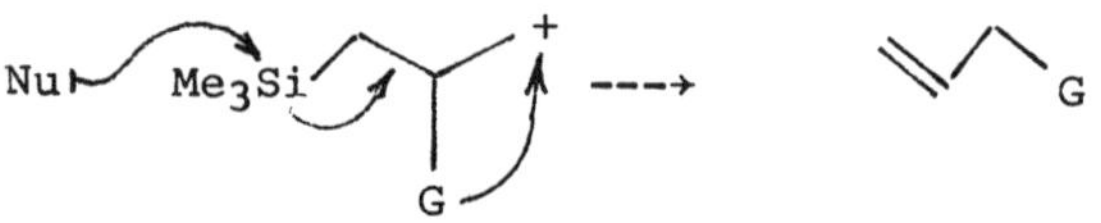

$$Nu \frown Me_3Si \frown \overset{+}{\frown}_G \longrightarrow \diagdown\!\!\diagup\!\!\diagdown_G \qquad (10)$$

is the ease of removal of the electrofugal silyl group by a nucleophile and/or the ability of the silyl group to stabilize the rearranged cation by its large electron-donating effect that could be both +I and +M in origin.

<u>Base-Catalyzed Cleavage of Benzyl-Silicon Bond in Alcohol.</u> The cleavage of benzyltrimethylsilanes in NaOMe/MeOH involves a separation of the PhCH$_2^-$ carbanion in the rate-determining step with MeO$^-$ attacking the silicon either in a synchronous process (with the MeO-Si bond formed in the transition state) or in a prior step.

The PhCH$_2$SiMe$_2$(CH$_2$)$_m$OH compounds with m = 2 or 3 are cleaved 0.75 and 95-135 times faster, respectively, than PhCH$_2$SiMe$_3$ by NaOMe/MeOH at 50° [60]. This higher reactivity provides support for an intramolecular attack of the alkoxide center on the silicon in the cyclic transition state

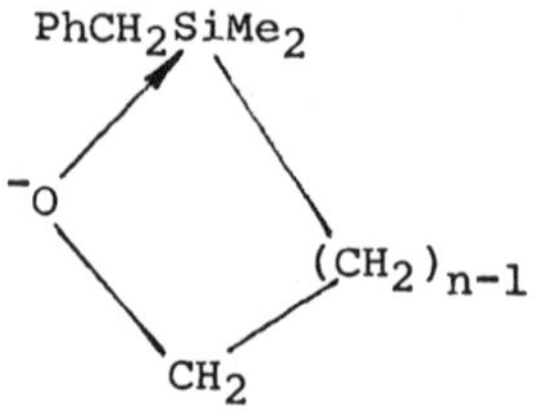

$$\begin{array}{c} PhCH_2SiMe_2 \\ \diagup \qquad \diagdown \\ ^-O \qquad\qquad (CH_2)_{n-1} \\ \diagdown \qquad \diagup \\ CH_2 \end{array}$$

and provides anchimeric assistance to the cleavage of the benzyl group.

Acetolysis of Alkyl Sulfonates. The acetolysis of ester derivatives of sulfonic acids, $ROSO_2R$, proceeds via formation of the intimate and the solvent-separated ion-pair, the reaction products being formed by reaction of these intermediates with acetic acid. During the heterolysis of the C_R-O bond, some degree of positive charge is generated at the carbon that bears the nucleofugal OSO_2R, and the rate-controlling step thus could accomodate both S_N1 and S_N2 as well as borderline behavior.

Acetolysis of alkyl methanesulfonates is accelerated by the introduction of a Me_3Si group into the γ-position. The small rate enhancement [61] for $Me_3Si(CH_2)_3CMs$ (k_{rel} = 14.0) relative to n-PrOMs (k_{rel} = 1) can be attributed to a greater stabilization of a partially positive intermediate by the weak +I effect of $Me_3Si(CH_2)_3$ group.

A similar rate accelerating effect of the remote silicon can be seen from the acetolysis of isosteric carbon and silicon cyclohexyl tosylates [62] (Table XVII). It is assumed that distortions of the cyclohexane ring increase the solvolysis rate by destabilizing the ground state relative to the transition state, i.e., both the cyclopentyl and the cycloheptyl tosylates solvolyze faster than the cyclohexyl [63]; thus, a three-to-five-fold increase in the rate of acetolysis in the silacyclohexyl system is minute when it is noted that cycloheptyl tosylate solvolyzes thirty times faster [63] than cyclohexyl tosylate, and this increase is likely to be associated with conformational effects alone.

Hydrolysis of Alkyl Acetates

Alkaline Hydrolysis. The rate-determining step of the alkaline hydrolysis of primary alkyl esters is assumed to be the attack of a hydroxyl anion on the carbonyl carbon of the ester. The ease of formation of the transition state depends on the extent of the electron deficiency and on the steric shielding of the reaction center. Examination of Figure 2, where the reactivities of n-alkyl acetates, $Me(CH_2)_mOC(O)Me$ (m = 2-8, 11), the $Me_3C(CH_2)_mOC(O)Me$ (m = 1-4), and the $Me_3Si(CH_2)_mOC(O)Me$ (m = 1-6) compounds are compared, shows that reactivities of silicon compounds

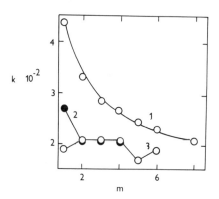

Fig. 2. Dependence [64] of rate constants of alkaline hydrolysis (L mol^{-1} sec^{-1} in 70% aqueous dioxane at 20°) of the compounds Me(CH$_2$)$_m$OC(O)Me (1), Me$_3$C(CH$_2$)$_m$OC(O)Me (2) and Me$_3$Si(CH$_2$)$_m$OC(O)Me (3) on the number (m) of methylene groups.

are the lowest. The initially (n = 0-2) larger and the later (n = 3-11) smaller decrease in reactivity of n-alkyl acetates is likely caused by an enveloping of the carbonyl group by the alkoxy O(CH$_2$)$_m$Me group, which results both in steric shielding of the reaction center and also in steric hindrance toward solvation of the carbonyl group. The relatively greater decrease in reactivity of the Me$_3$C(CH$_2$)$_m$OC(O)Me series can be attributed to a more effective shielding of the carbonyl carbon by the Me$_3$C(CH$_2$)$_m$O group. The reactivity of Me$_3$Si(CH$_2$)$_m$OC(O)Me is always slightly less than that of the appropriate Me(CH$_2$)$_m$OC(O)Me or Me$_3$C(CH$_2$)$_m$OC(O)Me compound and cannot be associated with the fact that the steric effect of the Me$_3$Si group is greater than that of the Me$_3$C group. The explanation can be given that the Me$_3$SiCH$_2$ group (m = 1) deactivates the carbonyl group by a weak +I effect, and, when m is higher, a weak (p-d)$_\sigma$ O→Si interaction

INTRAMOLECULAR INTERACTION

Table XVII. Rates [62] of acetolysis of cyclohexyl tosylates at 70°C

Tosylate	$k \cdot 10^4$, sec^{-1}	Tosylate	$k \cdot 10^4$, sec^{-1}
 TsO–[cyclohexyl]–Si(Me)(Me), H	1.05	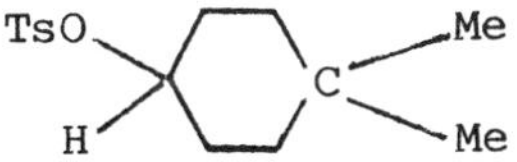 TsO–[cyclohexyl]–C(Me)(Me), H	0.28
 TsO–[cyclohexyl]–Si(Ph)(Ph), H	0.31	—	—
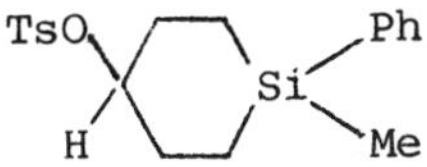 TsO–[cyclohexyl]–Si(Ph)(Me), H	0.61	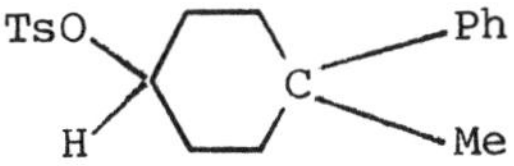 TsO–[cyclohexyl]–C(Ph)(Me), H	0.20
 TsO–[cyclohexyl]–Si(Me)(Ph), H	0.59	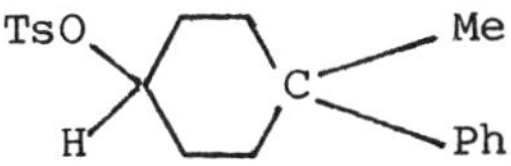 TsO–[cyclohexyl]–C(Me)(Ph), H	0.14

[Diagram: Me—C(=O)—O ⇠ Me₃Si—(CH₂)ₘ]

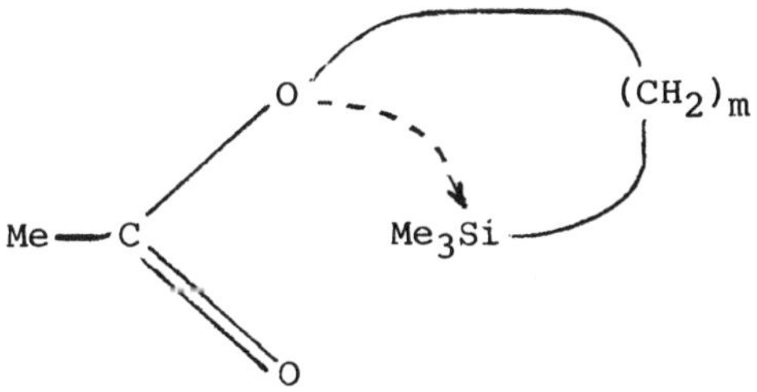

brings the Me$_3$Si group into close proximity to the reaction center.

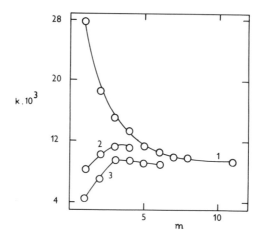

Fig. 3. Dependence [65] of rate constants (L mol^{-1} sec^{-1} in 70% aqueous dioxane at 65°) of acid hydrolysis of the acetates Me(CH$_2$)$_m$OC(O)Me (1), Me$_3$C(CH$_2$)$_m$OC(O)Me (2), and Me$_3$Si(CH$_2$)$_m$OC(O)Me (3) on the number (m) of methylene groups.

<u>Acid Hydrolysis</u>. The rate-determining step of acid hydrolysis of primary esters is the formation of a complex by the nucleophilic attack of a H$_2$O molecule on the protonated form of the ester. The ability of the oxygen of the RO group to transfer the electronic effect of the substituent, R, on to the reaction center is weaker in this reaction than in alkaline hydrolysis.

The rates of acid hydrolysis of acetates such as L(CH$_2$)$_m$OC(O)Me with L = Me, Me$_3$C and Me$_3$Si (Fig. 3) show [65] a somewhat similar dependence on the number of methylene groups, as do the rates of alkaline hydrolysis. The decrease in rate with higher m can again be ascribed to a more effective enveloping of the reaction center by the (CH$_2$)$_m$R group. Contrary to the situation with alkaline hydrolysis, the reactivities of Me$_3$Si(CH$_2$)$_m$OC(O)Me (m = 3, 4) are not lower than those of their carbon analogues but have the same value. This may indicate that a weak O→Si coordination bond increases the steric effect of the (CH$_2$)$_m$SiMe$_3$ group over that of the (CH$_2$)$_m$CMe$_3$ group during alkaline hydrolysis of Me$_3$Si(CH$_2$)$_m$OC(O)Me compounds but is

overcome by the stronger requirements of the electrophile
during acid hydrolysis of these compounds:

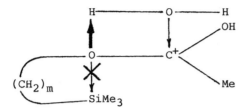

Hydrolysis of Acetals. The acid hydrolysis of acetals is governed by the rate of the unimolecular breakdown of the conjugate acid of the acetal (Eq. 11).

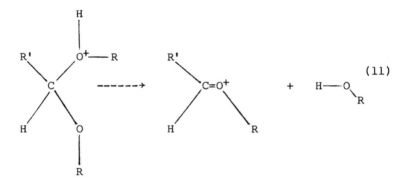

A trialkylsilyl group introduced into the γ-position shows [66] a deactivating effect toward acid-catalyzed hydrolysis. The reactivity of acetals increases in the following order,

$$\underset{O(CH_2)_3SiR_3}{\overset{O(CH_2)_3SiR_3}{R'HC}} < \underset{OR}{\overset{O(CH_2)_3SiR_3}{R'HC}} < \underset{OR}{\overset{OR}{R'HC}}$$

which can be associated with stabilization of the oxonium ion by a +I effect of the $Me_3Si(CH_2)_3$ group.

Hydrolysis of Alkyl Vinyl Ethers. The acid-catalyzed hydrolysis of alkyl vinyl ethers is a reaction in which the rate-determining step consists of a proton transfer to vinyl ether (Eq. 12).

$$H_2C=CH-OR \xrightarrow{H^+} Me-CH=O^+-OR \qquad (12)$$

Trialkylsilyl groups, when placed in the γ-position, have little effect on the reaction rate; there is no difference in the rates of substituted Me_3Si- and Et_3Si- derivatives (Table XVIII). The substitution at the β-carbon results in an increase of the rate, and it is presumed [67] that the rate-accelerating effect of the β-Me_3Si group is associated with its electron-donating ability. The higher reactivity of $Me_3Si(CH_2)_2OCH=CH_2$, compared to $Et_3Si(CH_2)_2OCH=CH_2$, is apparently associated with steric factors.

<u>Methanolysis of Trimethylsilyl Ethers</u>. The base-catalyzed methanolysis of trimethylsiloxyalkanes includes synchronous bond-making and bond-breaking in the transition state.

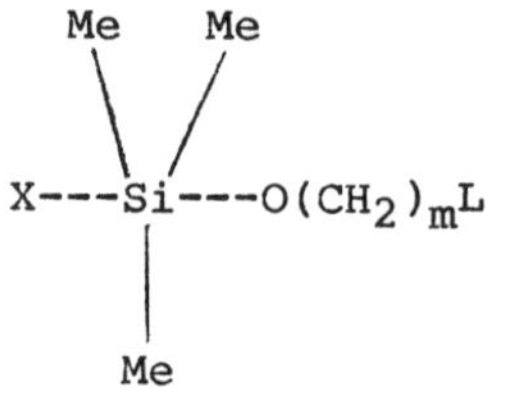

```
        Me   Me
         \  /
          \/
   X---Si---O(CH₂)ₘL          (X = MeO•••HB, or B)
          |
          Me
```

Table XVIII. Rates [67] of acid-catalyzed hydrolysis of $LOCH=CH_2$ at 25°C

L	$k \cdot 10^2$, $L \cdot mol^{-1} \cdot sec^{-1}$	$\Delta H^{\ddagger}$, $kJ \cdot mol^{-1}$	$\Delta S^{\ddagger}$, e.u.
$Me_3Si(CH_2)_2$	7.8	50	24.5
$Et_3Si(CH_2)_2$	7.0	68	9.5
$Me_3Si(CH_2)_3$	4.9	68	10.2
$Et_3Si(CH_2)_3$	4.9	67	10.6

The relative rate constants of the ethylamine (B)-catalyzed methanolysis of $L(CH_2)_mOSiMe_3$ (L = Me, Me_3C and Me_3Si) are given in Table XIX and follow [68] the extended Taft equation (Eq. 13).

$$\log k_{rel} = \log(k^{R^xOSiMe_3}/k^{EtOSiMe_3}) =$$
$$\rho^*(\sigma^* + 0.10) + \delta(E_s + 0.07) + C \qquad (13)$$

where $\rho^* = 4.64$, $\delta = 0.62$ and $C = -0.07$.

The reactivities of $L(CH_2)_mOSiMe_3$ for L = Me, Me_3C, and Me_3Si are comparable from m = 2 to higher values of m (Table XIX). The high reactivity of $Me_3SiCH_2OSiMe_3$ in this reaction reflects a significant reduction of the +I effect of the Me_3SiCH_2 group in this compound. This view is supported [68] by Eq. (13), on the basis of which the E_s parameter for the Me_3Si group, assuming σ^* for Me_3Si equal to -0.26, was calculated to be -0.09. This unlikely value indicates an intramolecular interaction between the oxygen and silicon atoms that would lead to a decrease of the electron density on the silicon of the Me_3SiO group:

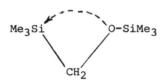

Table XIX. Relative rate constants of base-catalyzed methanolysis of $L(CH_2)_mOSiMe_3$ compounds [68]

L	m					
	1	2	3	4	5	6
Me	1.0	0.60	0.53	0.54	0.52	0.53
Me_3C	0.05	0.52	0.52	0.51	-	-
Me_3Si	0.15	0.49	0.50	0.52	0.50	0.49

Acid-Base Equilibria of Alcohols and Ethers. The relative base strengths (proton affinities) of ethers, ROR', can be determined by comparing the magnitudes of the infrared frequency shifts of the free and the hydrogen bonded proton donor (phenol, methanol, or pyrrole) on its complexation with ether. In a similar way, the acidities of alcohols are provided by the difference between the free and the hydrogen-bonded OH frequency of alcohols when they interact with the proper proton acceptor (tetrahydrofuran, dimethylsulfoxide). The difference between the free and hydrogen-bonded OH frequency, $\Delta \nu$, is directly related [69,70] to the basicity or acidity of the ether or alcohol, respectively, and can be correlated with polar effects of groups adjoining the oxygen [71].

The acidity of trimethylsilylalkyl alcohols, $Me_3Si(CH_2)_mOH$, reflects [55,72] the ability of the Me_3Si group to increase the electron density on the adjacent carbon atom. In the $Me_3Si(CH_2)_mOH$ series, trimethylsilylmethanol is the least acidic. The acidity of $Me_3Si(CH_2)_2OH$ is higher than that of methanol but lower than that of other silylalkylalcohols in which the +I effect of the Me_3Si group is suppressed by three or four methylene groups (Table XX).

The relative basicity of silyl substituted ethers, $X_nR_{3-n}Si(CH_2)_mOR$, is influenced by the silyl group only when this group is separated from the oxygen by less than two CH_2 units. Table XXI shows that the oxygen basicity is increased by the Me_3Si group but is decreased by the Ph_2MeSi, Ph_3Si, H_3Si, and Cl_3Si groups.

Table XX. Acidity of $Me_3Si(CH_2)_mOH$ as $\Delta \nu(OH)$, in cm^{-1}, upon complexation with tetrahydrofuran and acetonitrile [55]

Proton acceptor	m			
	1	2	3	4
Tetrahydrofuran	145	149	155	154
Acetonitrile	91	95	99	98

Table XXI. Relative basicity, $\Delta\nu OH$ in cm^{-1}, of some silyl-alkyl ethers [73-76]

Series	Pronton donor	m=1	m=2	m=3	m=4
$Me_3Si(CH_2)_mOSiMe_3$	phenol	293	289	277	278
$Me_3Si(CH_2)_mOMe$	pyrrole	143	150	132	-
$Me_3Si(CH_2)_mOH$	phenol	246	240	233	232
$Ph_2MeSi(CH_2)_mOMe$	methanol	127	145	138	138
$Ph_3Si(CH_2)_mOMe$	methanol	114	142	141	-
$Cl_3Si(CH_2)_mOMe$	pyrrole	76	98	128	-

The Me and H_3Si groups are slightly electron-releasing when attached to a carbon atom, and their electronegativities are nearly equal [77]. Contrary to these electronegativity considerations, the relative basicity ($\Delta\nu(OH)$ of phenol) of H_3SiCH_2OMe (246 cm^{-1}) is lower [78] than that of EtOMe (268 cm^{-1}) and indicates that the H_3SiCH_2 group is electron-withdrawing. It is understood [78] that intramolecular interaction between the oxygen and silicon takes place in H_3SiCH_2OMe and results in a decrease of electron density at the oxygen. A similar but weaker intramolecular interaction cannot be ruled out in other silylmethyl ethers [78].

The polar effect parameters σ^* of silyl groups in methyl ethers are calculated on the basis of a linear plot between σ_{RX}^* and $\Delta\nu$ (OH) for R^X-OMe compounds and the equation interrelating [79,80] $\Delta\nu$ (OH) of methanol or $\Delta\nu$ (NH) of pyrrole to $\Delta\nu$ (OH) of phenol, and are gathered in Table XXII.

2.2.7 Reactions of Thiols and Sulfides

The ability of a sulfur atom to act as a reaction center in reactions of sulfur-containing silanes can be influenced by the electron-releasing +I effect of $Me_3Si(CH_2)_m$ groups in two ways: in addition reactions of thiols the sulfur atom is activated; in homolytic processes the sulfur atom is deactivated.

<u>Heterolytic Additions of Thiols</u>. The amine catalyzed addition of thiols to phenyl isocyanate very likely proceeds [81] in two steps, A and B (Eq. 14).

$$RSH + R_3\overset{..}{N} \underset{A}{\rightleftarrows} RS\cdots H\cdots NR_3^- \underset{PhNCO}{\overset{B}{\dashrightarrow}} PhNHC(O)SR + NR_3^- \quad (14)$$

The second rate-determining step involves a nucleophilic attack on PhNCO by the sulfur atom, the nucleophilicity of the sulfur being the predominant factor. The reactivity of $Me_3Si(CH_2)_mSH$ (m+k_{rel}: 1-1.37; 2, 3, 4-1) decreases with the decreasing influence of +I effect of the Me_3Si group. The small increase in the reactivity of Me_3SiCH_2SH and the reactivity of $(Me_3SiO)_{3-n}Me_nSiCH_2SH$ (n-k_{rel}: 0-1, 1-1.7, 2-1.7) is consistent [81] with equilibrium A (Eq. 14) leveling the effect of the substituents.

Table XXII. Taft σ^* parameters of silylmethyl groups in silylmethyl ethers

Silyl group	σ^*
Me_3SiCH_2	-0.19
$PhMe_2SiCH_2$	0.11
H_3SiCH_2	0.27
Ph_3SiCH_2	0.32
Cl_3SiCH_2	0.93

A similar order of reactivity is observed [82] for the KOH catalyzed addition of $R_3Si(CH_2)_mSH$ compounds to acetylene; the rate of this reaction is also controlled by the nucleophility of the attacking species, which in this case is the $Me_3Si(CH_2)_mS^-$ anion.

Homolytic Reactions of Sulfides. The radical addition of silylalkylthiols to vinylthioalkylsilanes (Eq. 15) is facilitated by an increased distance between the silicon and sulfur atoms.

$$R_3Si(CH_2)_mCH=CH_2 + Me_3Si(CH_2)_3SH \dashrightarrow$$

$$R_3Si(CH_2)_mSCH_2CH_2S(CH_2)_3SiMe_3 \qquad (15)$$

The yields of products are lower than those formed in the addition of thiols to alkylvinylsulfides [82]. Since the reactivity of $R_3Si(CH_2)_mSCH=CH_2$ decreases in the order m = 3 > 2 > 1, trialkylsilyl substituents have a similar deactivating effect in the oxidation [83] of silylalkyl vinyl sulfides by hydrogen peroxide (Eq. 16).

$$R_3Si(CH_2)_mSCH=CH_2 \xrightarrow{H_2O_2} R_3Si(CH_2)_mS(O)CH=CH_2 \qquad (16)$$

Both reactions demonstrate that an increase of electron density on sulfur impedes the ability of $CH_2=CH-SR$ molecules to react in homolytic processes.

2.2.8 Reactions of Alkyl Halides

The mutual effect of silicon and the carbon-functional group can be investigated either by means of the reactivities of the functional group itself and the reactivity of the functional group-carbon bond, or by means of the reactivity of other molecular centers. The reactions discussed below are therefore separated into (a) reactions involving change on the center bearing the halogen, and (b) reactions occurring at other molecular centers. The latter reactions aid in understanding the effect of silyl groups on the rate of reactions involving a change of the C-halogen bond. They reflect a strong reduction in the electron-withdrawing ability of YCH_2 groups attached to silicon and also that the molecular parameters of (halogenomethyl)silanes are not controlled by a purely inductive electron release (Si→C) but also by an intramolecular interaction between halogen and silicon, as depicted below:

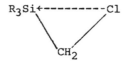

Carbonium-Ion Forming Reactions

S_N1-Like Substitution. The S_N1-like substitution of alkyl halides with silver nitrate in ethanol is controlled by the ionization of alkyl halides that is simultaneously assisted [84] by both the electrophilic $AgNO_2$ and the nucleophilic NO_3^-:

$$NO_3^- \cdots R-X \cdots Ag^+NO_2^-$$

These reactions occur even with primary halides branched in the neighborhood of the reaction center, but they are completely inhibited by alkylhalides substituted in the α-position by the Me_3Si group. Thus, Me_3SiCH_2Y (Y = Cl, Br) compounds do not react [85] with silver nitrite in ethanol under conditions under which the corresponding n-hexyl halides or Me_3CCH_2Y compounds react quite readily [84], although the +I effect of Me_3Si group should assist this reaction by stabilizing the forming carbonium ion.

A similar deactivating effect of the Me_3Si group(s) in the α-position to the reaction center in alkylhalides is exerted in solvolytic reactions of α-(trimethylsilyl)alkyl halides with aqueous acetone or ethanol. No detectable reaction ($k_1 < 3 \cdot 10^{-7}$ sec^{-1}) was observed [86] with $(Me_3Si)_nCH_{3-n}Y$ and Me_3SiCMe_2Y compounds (Y = Cl, Br) under conditions when a unimoleuclar process with Me_3CY and Me_3CCMe_2Y halides proceeds rapidly. The decrease in the ease of reaction by an ionization mechanism in the case of substitution of a Me_3Si group in place of either a Me or a Me_3C group cannot be explained by steric effects, since the Me_3C group offers an even greater hindrance.

Olefin Formation. The solvolysis of Me_3SiMe_2CBr in aqueous ethanol proceeds some 38,000 times slower [87]

INTRAMOLECULAR INTERACTION

than the reaction of its analogue, Me_3CMe_2CBr, and gives isopropenylsilane. The ρ value (-1.1) obtained from the correlation of the rates of solvolysis of the aryl derivatives, $XC_6H_4SiMe_2CMe_2Br$, with Hammett σ constants is consistent with a substantial amount of positive charge at the carbon adjacent to silicon in the transition state (believed [88] to be represented by the structure below).

$$\begin{array}{c} \text{H} \\ H_2C\text{---}H\text{----}OEt \\ | \\ RMe_2Si\text{------}C^{\delta+}\text{--------}Br^{\delta-} \\ | \\ Me \end{array}$$

If so, the reactions of Me_3CMe_2CBr and Me_3SiMe_2CBr are substantially identical in mechanism and might indicate that the Me_3Si group is significantly poorer at stabilizing an adjacent carbonium-ion center than is a t-butyl group. The kinetic data can also be explained, however, by a ground state intramolecular interaction between the silicon and halogen that, as will be shown in the following section, is responsible in (halogenomethyl)silanes for a decreased electron density on halogen.

The solvolysis of 2-(halogenoalkyl)silanes, $R_3SiCH_2CH_2Y$, is roughly as ionic in character as is the solvolysis of t-butyl chloride, its rates being substantially greater than those of isopropyl and t-butyl chlorides [2]. After solvolysis of $Me_3SiCH_2CD_2Br$ by aqueous MeOH is allowed to proceed to about 50% completion, the recovered bromide contains $Me_3SiCH_2CD_2Br$ and $Me_3SiCD_2CH_2Br$ in a 1:2 ratio [89]. The migration of the Me_3Si group during the solvolysis is consistent with a mechanism [89] involving an anchimerically assisted ionization of the C-X bond to give the silacyclopropenium ion (Eq. 17).

$$Me_3SiCH_2CD_2Br \rightleftharpoons \underset{\underset{CH_2=CD_2 + Me_3SiBr}{\downarrow}}{\overset{SiMe_3}{\underset{+}{\triangle}}CH_2\text{---}CD_2} + Br^- \rightleftharpoons BrCH_2CD_2SiMe_3 \quad (17)$$

A similar mechanistic pathway can be written for the highly stereospecific trans-elimination of trimethylbromosilane during solvolysis of erythro-1,2-dibromopropyltrimethylsilane. It was suggested [56] that this reaction involved both anchimerically assisted trans-elimination and non-stereospecific elimination via a classical carbonium ion (Eq. 18). The latter pathway is favored in more highly ionic solvents.

S_N2 Displacement Reactions. These reactions differ from carbonium-ion forming reactions by a smaller development of positive charge in the transition state. For saturated halides, electron-withdrawing substituents usually decrease S_N2 reactivity, and electron-donating substituents produce rate increases, but both effects are fairly small. The effect of silyl groups upon the rate of

$$\text{(Newman projection with Br, H, H, SiMe}_3\text{, Br, Me)} \longrightarrow \overset{\text{SiMe}_3}{\underset{\text{Br}}{\text{C}}}\overset{\text{H}}{\underset{\text{Me}}{-}}\overset{\text{H}}{\underset{\text{Me}}{\text{C}}} \longrightarrow \overset{\text{H}}{\underset{\text{Br}}{\text{C}}}=\overset{\text{H}}{\underset{\text{Me}}{\text{C}}} \quad (18)$$

$$\updownarrow$$

$$\overset{+}{Me_3SiCHBr-CHMe} \longrightarrow CHBr=\!\!=\!\!CHMe$$

alkyl halides in S_N2 reactions is not uniform and depends upon the character of the solvent.

In aprotic solvents such as acetone, Me_3SiCH_2Cl is considerably (~15 times) more reactive [88,90] than n-butyl chloride; the activation enthalpy for the reaction of Me_3SiCH_2Cl being lower by 16.3 kJ/mol, and the activation entropy higher by 6.3 e.u., than for the reaction of n-BuCl. These parameters [91] are consistent with a considerable amount of hindrance of the Me_3Si group and its strong activation of the adjacent carbon. The activating effect of the Me_3Si group does not arise from electrons being supplied from silicon to the central carbon atom, since increasing the electron release of the organosilyl group in p-L-$C_6H_4SiMe_2CH_2Cl$ compounds results in a decrease [90] of the reactivity of these compounds in the reaction with iodide ion:

L:	Cl	H	Me	MeO
k_2 (h^{-1}·mole^{-1}):	2.58	1.13	0.88	0.87

The activating effect of the silyl group was therefore attributed [85,90] to the stabilization of the transition state by interaction of the incoming nucleophile ion with the vacant d orbitals of the silicon atom as well as with the carbon nucleus (Eq. 19). Such a bridging interaction can then be suggested to operate also in the reaction of Me$_3$SiCH$_2$Cl with thiocyanate [92,93], since this reaction is faster than that of Me$_3$Si(CH$_2$)$_3$Cl.

The reactivity of Me$_3$SiCH$_2$Cl in S$_N$2 reactions

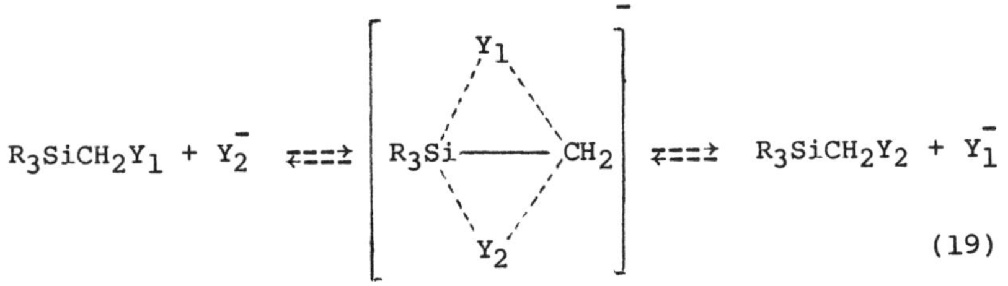

$$R_3SiCH_2Y_1 + Y_2^- \rightleftharpoons \left[R_3Si \underset{Y_2}{\overset{Y_1}{\diamond}} CH_2 \right]^- \rightleftharpoons R_3SiCH_2Y_2 + Y_1^- \quad (19)$$

decreases in protic solvents, but an activating effect of silicon cannot be utterly ruled out. Thus, this effect probably operates in the reaction with alkoxide ions in alcohols, although Me$_3$SiCH$_2$Cl is somewhat (2-4 times) less reactive than n-BuCl [94] or Me$_3$Si(CH$_2$)$_3$Cl [95], but this difference in reactivity is much smaller than would be anticipated on the basis of pure steric hindrance of the Me$_3$Si group. Only a slightly lower reactivity of Me$_3$SiCH$_2$Cl compared to the n-hexyl compound was found for reactions that do not proceed at all with Me$_3$CCH$_2$Cl (Table XXIII).

In aqueous ethanol, Me$_3$SiCH$_2$I undergoes [96] iodide exchange (Eq. 20) less readily than does EtI:

$$LI + NaI^* \rightleftharpoons LI^* + NaI \quad (20)$$

Table XXIII. $S_N 2$ reactions of Me_3SiCH_2Cl and $n-C_6H_{13}Cl$ in protic media [95]

Reagent	Solvent	Percentage of reaction	
		Me_3SiCH_2Cl	$n-C_6H_{13}Cl$
KOC(O)Me	EtOH	23	29
KOH	EtOH	52	69
KOH	HOC(O)Me	26	32
KOH	H_2O-EtOH	16	22

L	Et	n-Pr	Me_3SiCH_2	Et_3SiCH_2
$k_2 \cdot 10^5$ (mol^{-1}·L·sec^{-1})	28	13	6.0	2.0

The explanation was offered [86] that the more effective solvation of both nucleophile and nucleofug in this medium both sterically hinders bridging and also reduces the need for it, and in such a case the Me_3Si group should affect the adjacent carbon center with a rather poor stabilizing effect such as is observed in carbonium ion forming reactions. Such an explanation, however, is not consistent with the fact that the large polarizable iodide ion is solvated by protic alcohol less than with aprotic acetone. It is therefore tempting to assume that the increased reactivity of silylmethyl halides in $S_N 2$ reactions is associated with the lack of electron density on the $-CH_2-$ carbon due to a ground state intramolecular Cl→Si interaction, and that its ability to survive in the transition state of $S_N 2$ reactions depends on the nature of the nucleophile and its solvation. Then, two different mechanisms could be written for two extreme cases (Eq. 21), which express a higher probability of reaching the transition state with a bridging interaction with better solvated or with less solvated but more polarizable nucleophiles.

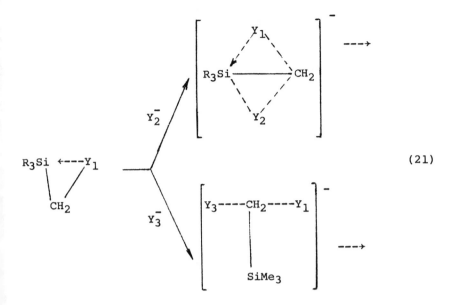

(21)

Reaction with Sodium Atoms. In the reaction of an organic halide with a sodium atom, the latter approaches the halide until its valence electron is transferred to the halogen; the organic halide then dissociates. The reaction is controlled by a dissociative splitting of the C-Cl bond, its ease increasing [97] with a decrease of negative charge on the halogen. The following bimolecular rate constants ($k = 10^{11}$ mL mol^{-1} sec^{-1}; 520 K)

Me_3SiCH_2Cl	16.2	Me_3CCH_2Cl	1.06
$(Me_3Si)_2CHCl$	230	$Me_3C(CH_2)_2Cl$	1.39
$Me_3Si(CH_2)_3Cl$	3.1		

reflect [98] the enhanced rates of α-silylsubstituted alkyl halides and can be rationalized in terms of a structure

$$Me_3Si\text{————}\overline{CH_2} \quad \boxed{Cl} \qquad Na^+$$

that is consistent with a reduction of electron density on the chlorine due to a ground state Cl→Si interaction. The rate-accelerating effect of silyl groups in this reaction thus originates in their electron-withdrawing ability.

Reaction With Grignard Reagents. In a reaction that features the exchange of fluorine in $Me_3Si(CH_2)_mF$ for the bromine of EtMgBr, the reactivity of silylalkyl fluorides decreases [99] with increasing distance between the silicon and fluorine (m, k_{rel}: 1, 33.8; 3, 5.56; 5, 1.0). The reaction may proceed via a four-center intermediate,

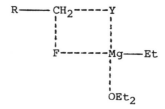

and it is not clear whether the reaction is controlled by nucleophilic attack of Y on the CH_2F group or by the ability of fluorine to coordinate with magnesium.

Hydrolysis of Organosilicon Hydrides. The rate-determining step of the acid-catalyzed hydrolysis of organosilicon hydrides, $X_nR_{3-n}Si-H$, consists of the formation of the transition state

$$[H_2O\overset{\delta+}{---}SiX_nR_{3-n}\overset{\delta-}{----}H\overset{\delta+}{---}H---OH_2]^+$$

by an electrophilic attack of a hydronium ion and a nucleophilic attack by a water molecule [100]. The rate is slightly increased by electron withdrawal from the silicon atom. The $ClCH_2SiMe_2$-H compound reacts much more slowly than would be anticipated from the correlation between the first-order rate constant and the Taft substituent parameter

$$\log k = \sum \sigma^*_{R,X} \cdot \rho^* + C$$

of R and X substituents. It thus appears [101] that not a steric effect but rather a reduced electron-withdrawing ability of the $ClCH_2$ group is responsible for the decreased reactivity of $ClCH_2SiMe_2H$. The σ^* parameter for the $ClCH_2$ group in this reaction approaches zero, the value known for the Me group.

Acid-Base Equilibria. The IR absorption spectra of the hydrogen bonds of phenol interacting with ethers in CCl_4 solvent reflect the proton-accepting ability of the oxygen of the ether [102]. The determination of oxygen basicities in aprotic CCl_4 should be only slightly influenced by solvation phenomena. It is based on the relationship [103] between the enthalpy of adduct formation (Eq. 22) and the change in frequency of the OH stretching vibration, $\Delta\nu$ of phenol upon complexation, the general applicability of which for appropriately chosen systems has been confirmed [104].

$$[R-O-R']_{CCl_4} + [PhOH]_{CCl_4} \rightleftarrows \left[R-O \begin{smallmatrix} \cdots H \cdots O-Ph \\ \\ R' \end{smallmatrix} \right]_{CCl_4} \quad (22)$$

An examination of the oxygen basicity of $(EtO)_n Me_{3-n} SiCH_2 Y$ compounds provides information on the polar effect of the YCH_2 group, since a linear relationship (Eq. 23) between the $\Delta\nu$ of phenol and the Taft parameter $\sum \sigma^*_{R,R^X}$

$$\Delta\nu(OH) = \rho^* \sum \sigma^*_{R,R^X} + C \quad (23)$$

for (R or $R^X)_n Si(OEt)_{4-n}$ with R or R^X exerting a purely inductive effect has been demonstrated [105]. The polar effect parameter σ^* for $CH_2 Y$ groups in the above compounds, given in Table XXIV, shows that the electron-releasing action of these groups is reduced with regard to their known -I effect in organic compounds. Their lower electron-withdrawing effect is associated with an intramolecular interaction between the silicon and the halogen, which leads to electron transfer from the latter to the former atom [108,109].

Friedel-Crafts Alkylation. The key step in this aromatic substitution is the heterolysis of the C-Cl bond of an alkyl chloride, which is assisted by the electron deficient metal halide.

The reactivity of alkyl chlorides substituted by $Cl_n Me_{3-n} Si$ groups diminishes [110] with decreasing distance between the chlorosilyl group and the C-Cl bond (Table

XXV). The rates show that Cl_3Si and Cl_2MeSi groups affect the C-Cl bond by their electron-withdrawing -I effect and decrease the polarizability that is so important for the facilitation of the alkyl chloride-$AlCl_3$ interaction. The low value of the reaction rate for n = 2, m = 3 and other values must be treated with caution, however, since reactivities were determined from the rates of evolution of hydrogen chloride and not from the rates of formation of the products.

2.3 REACTIONS WITH PATHWAY DOMINATED BY THROUGH-SPACE INTERACTION BETWEEN SILICON AND A FUNCTIONAL GROUP

Through-space interactions between the silicon and a center of the functional group that occur in the ground

Table XXIV. The σ^* parameter of YCH_2 groups from $\Delta\nu$ (OH) of phenol [106] upon complexation with $(EtO)_nMe_{3-n}SiCH_2Y$

Y	n	σ^*YCH_2	σ^*YCH_2 in org. compds. [107]
Cl	1	0.57	
Cl	2	0.51	1.05
Cl	3	0.59	
Br	1	0.68	
Br	2	0.51	1.00
Br	3	0.59	
I	1	0.45	
I	2	0.44	0.85
I	3	0.53	

Table XXV. Relative rates of $AlCl_3$-catalyzed alkylation [110] of benzene by alkyl chlorides $Cl_n Me_{3-n} Si(CH_2)_m Cl$

n	m	k_{rel}
3	1	1
3	2	18
3	3	23
2	1	1.6
2	2	189
2	3	28

state of a carbon-functional silane or during its reaction are designated as ℓ,m, which means that the centers in question are separated by $m-\ell$ atoms:

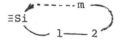

2.3.1 Reactions Involving 1,2-Interaction

<u>Interaction With Carbon</u>. Cyclopentadienyl- and indenylsilanes are compounds in which a weakly σ-bonded silyl group undergoes concerted intramolecular migration by either 1,2- or by random motions [111-114] (Eq. 24). For cyclopentadiene itself, the 1,2-hydrogen transfer reaction requires [115] 105 kJ/mol, compared to more than 167 kJ/mol postulated for the intramolecular transfer of a Me group in methylcyclopentadiene. In contrast, the transfer of a Me_3Si group in 5-(trimethylsilyl)cyclopentadiene requires [116] only 38 or less kJ/mol and leads to a very fast equilibration. Equilibrium measurements of the intra-

molecular transfer of hydrogen and the Me_3Si group with trimethylsilylcyclopentadiene revealed that, contrary to the carbon analogues, the 5-isomer is thermodynamically the most stable [117]. The rate of the silyl group

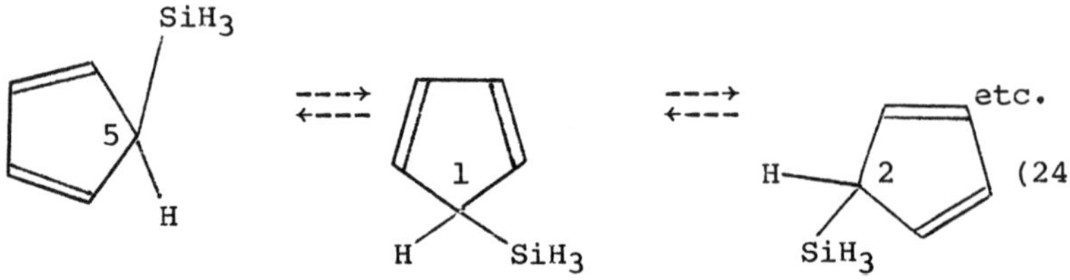

$$\underset{H}{\overset{SiH_3}{\diagdown}} \underset{5}{\bigcirc} \quad \xrightarrow{} \quad \underset{HSiH_3}{\bigcirc_1} \quad \xrightarrow{} \quad H-\underset{SiH_3}{\overset{}{\diagdown}}\underset{2}{\bigcirc}=\text{etc.} \quad (24)$$

rearrangement depends on the nature of the silyl group; it increases linearly [118] with successive substitutions of chlorine for methyl in the Me_3Si group. Most of the effect on the rates is due to changes in the entropy of activation. The rates of silyl group migration in various systems increase [119,120] in the order:

cyclononatetraene < indene < cyclopentadiene.

A thermally induced silyl group migration from carbon to vicinal carbon was also observed in cyclopropylsilanes [121,122] and pyrazolines [123], in which the migration terminus displays at least a partly unsaturated character. In these reactions there is the tendency for silicon to migrate in marked preference to any geminal hydrogen. With 5-trimethylsilylbicyclo[2.1.0]pentanes the trimethylsilyl group migration is 10^6 times faster [122] than that of hydrogen, and it occurs within a preformed 1,3 diradical (Eq. 25).

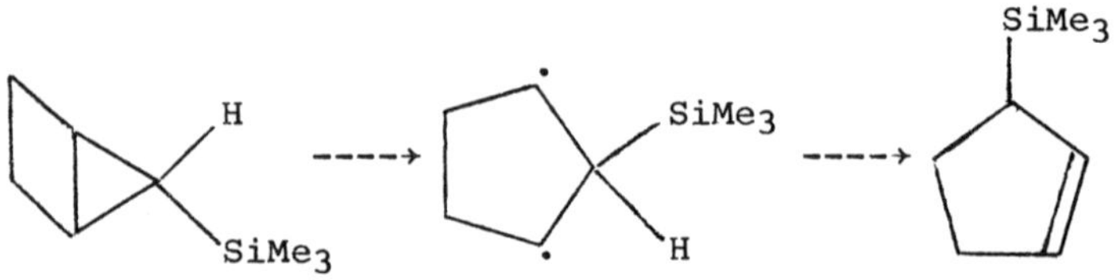

$$\text{structure} \quad \xrightarrow{} \quad \text{structure} \quad \xrightarrow{} \quad \text{structure} \quad (25)$$

In 3-carbomethoxy-4-trimethylsilyl-1-pyrazolines [123], the silyl group participates in the nitrogen expulsion, since 4-trimethylsilyl-substituted pyrazolines expel nitrogen at a rate approximately 10^7 faster than does 3-carbomethoxy-

3-methyl-1-pyrazoline. This reaction (Eq. 26)

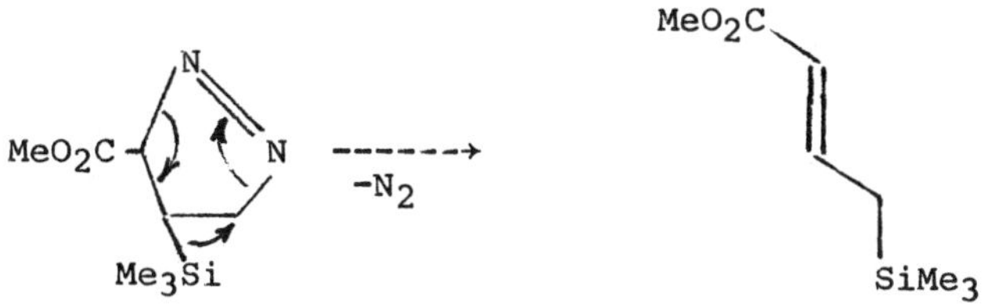

MeO2C
N
N
-N2
Me3Si
MeO2C
SiMe3

(26)

occurs by simultaneous reorganization of three σ bonds and bears a similarity to an intramolecular two σ-bond "dyotropic" rearrangement [124].

Interaction With Nitrogen. This migration takes place in the base-catalyzed rearrangement of secondary (aminomethyl)silanes [125,126]. The ionic mechanism proposed [126] involves an initial generation of nitrogen anion within which a 1,2-migration of silicon from carbon to nitrogen occurs, and, subsequently, an abstraction of a proton from the parent amine by the resulting carbanion (Eq. 27). The rearrangement is reversed, and the mechanism is consistent with the more rapid rearrangement of the benzylaminosilanes (L_1 = Ph) where stabilization of the carbanion relative to the nitrogen anion would occur, and also with the lack of rearrangement of the anilines (L_2 = Ph) where stabilization of the nitrogen anion relative to the carbanion should exist. A similar 1,2-anionic rearrangement of the alkyl group is unknown.

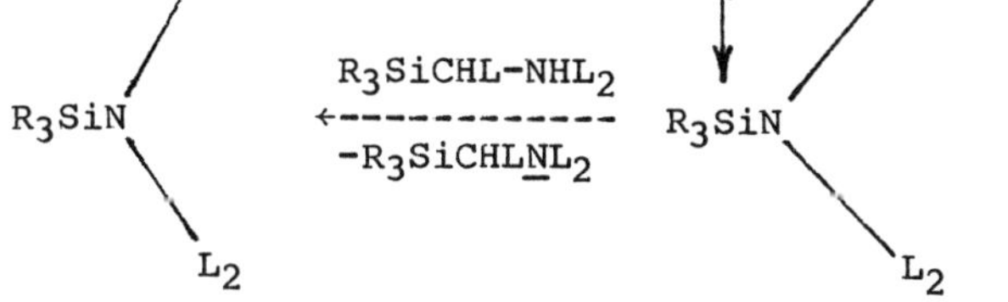

R3SiCHL1-NHL2
n-BuLi
-n-BuH
R3Si
NL2
CHL1
CH2L1
R3SiN
L2
R3SiCHL-NHL2
-R3SiCHLNL2
R3SiN
CHL1
L2

(27)

Similar but thermal 1,2-rearrangements of silicon are presumed to occur in the triazolines [127] and pyrazole

[128] that are transiently formed in the following reactions (Eq. 28 and 29), respectively).

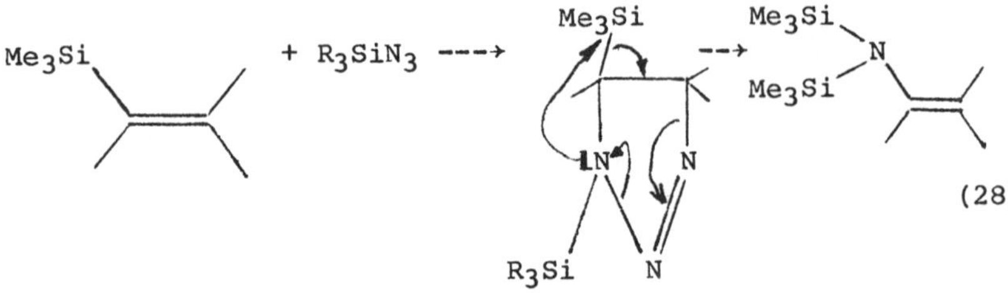

$$Me_3Si\diagup\diagdown + R_3SiN_3 \dashrightarrow [\text{intermediate}] \dashrightarrow Me_3Si-N(SiMe_3)-\text{vinyl} \quad (28)$$

$$MeO_2CC\equiv CCO_2Me + Me_3SiC(N_2)CO_2Et \dashrightarrow [\text{intermediate}] \rightarrow \text{pyrazole product} \quad (29)$$

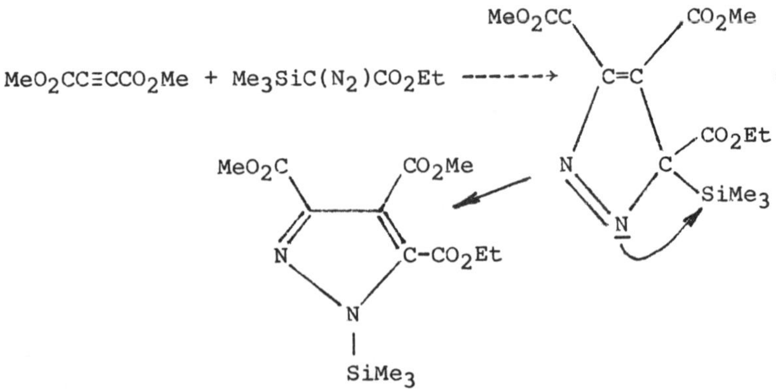

Interaction With Oxygen.

Silylcarbinols, $R_nPh_{3-n}SiCL_2OH$ (L = H, R, or Ph), rearrange [129] when treated with traces of proton-accepting reagents to the corresponding alkoxysilanes. The rates are very sensitive to the nature of the substituent on the carbinol carbon. Carbinols bearing phenyl groups rearrange 10^3 times faster than when alkyl groups or hydrogen are substituted for phenyl. The replacement of phenyl on silicon by methyl reduces the rates by only a factor of about six. The reaction proceeds as follows (Eq. 30).

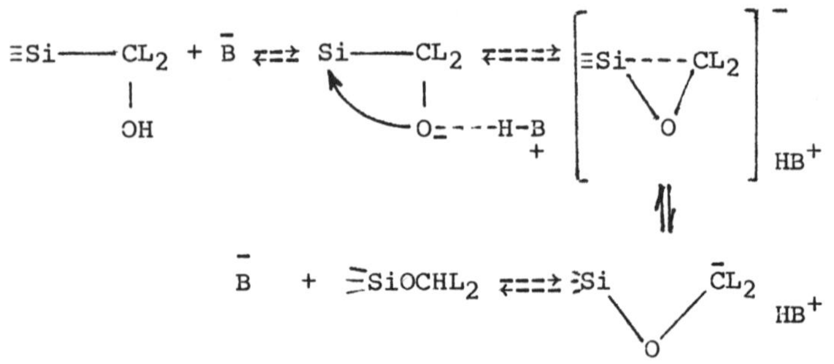

$$\equiv Si-CL_2(OH) + \bar{B} \rightleftharpoons Si-CL_2(O^--H-B) \rightleftharpoons [\equiv Si\cdots CL_2\cdots O]^- HB^+$$

$$\bar{B} + \equiv SiOCHL_2 \rightleftharpoons \equiv Si\diagdown O\diagup CL_2 \; HB^+ \quad (30)$$

INTRAMOLECULAR INTERACTION

An α-oxyanion is formed when the oxygen attacks at the silicon, which gives a carbanion via a three-center species with retention of configuration at the chiral silicon and inversion of configuration at the chiral carbon center. This rearrangement takes place in the course of reactions where an α-oxyanion is formed, namely during the reaction of ketones with nucleophilic reagents such as triphenylsilylpotassium [130] (Eq. 31), diazomethane [131] (Eq. 32), ethoxide ion [132] (Eq. 33), or alkylidenephosphoranes [133] (Eq. 34).

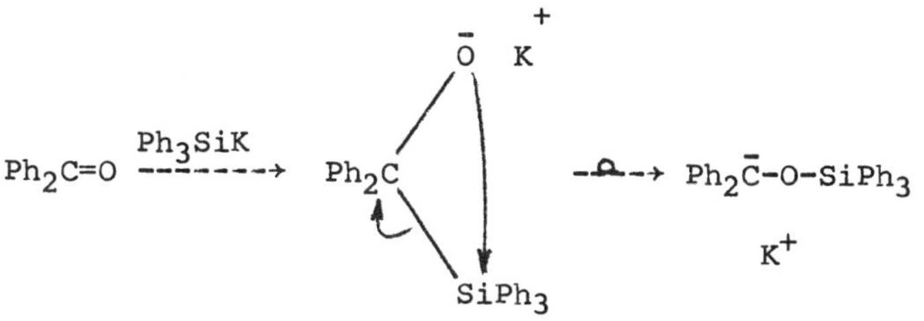

$$Ph_2C=O \xrightarrow{Ph_3SiK} Ph_2C\begin{matrix}\bar{O}\ \ ^+K\\ \diagup\ \ \diagdown\\ \ \ \ \ \ \ \ \ \ \ \\ \diagdown\ \ \diagup\\ SiPh_3\end{matrix} \longrightarrow Ph_2\bar{C}-O-SiPh_3 \quad K^+ \quad (31)$$

$$R_3SiCOR_1 \xrightarrow[-N_2]{CH_2N_2} R_3Si-\underset{CH_2^+}{\overset{\bar{O}}{\underset{|}{C}}}-R_1 \longrightarrow \underset{R_1}{\overset{R_3SiO}{\diagdown}}C=CH_2 \quad (32)$$

$$Ph_3Si-CO-Ph \xrightarrow{EtO^-} EtO-SiPh_2-\underset{|}{\overset{\bar{O}}{C}}Ph_2 \longrightarrow EtOSiPh_2-O\bar{C}Ph_2 \quad (33)$$

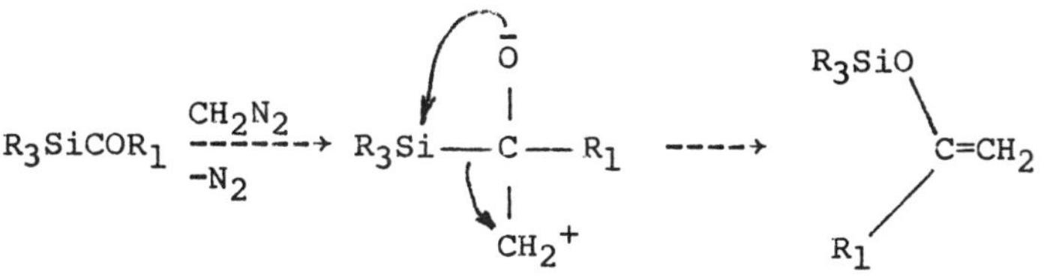

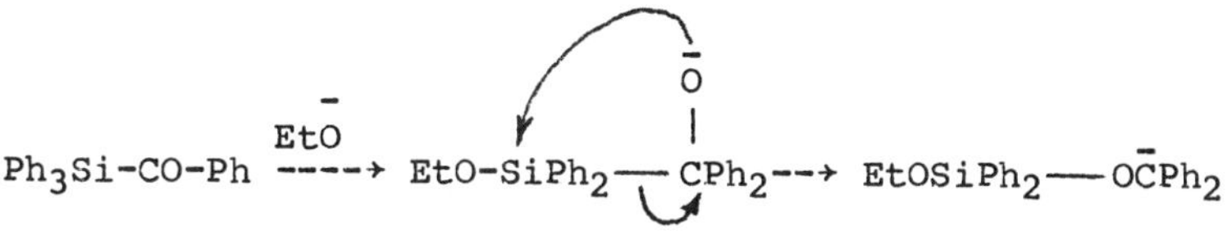

$$R_3SiCOPh \xrightarrow{Ph_3\overset{+}{P}-\overset{-}{CHR_1}} R_3Si-\underset{Ph}{\overset{\bar{O}\ \ \overset{+}{P}Ph_3}{\underset{|}{C}}}-CHR_1 \xrightarrow{-Ph_3P} R_3SiOCPh=CHR_1 \quad (34)$$

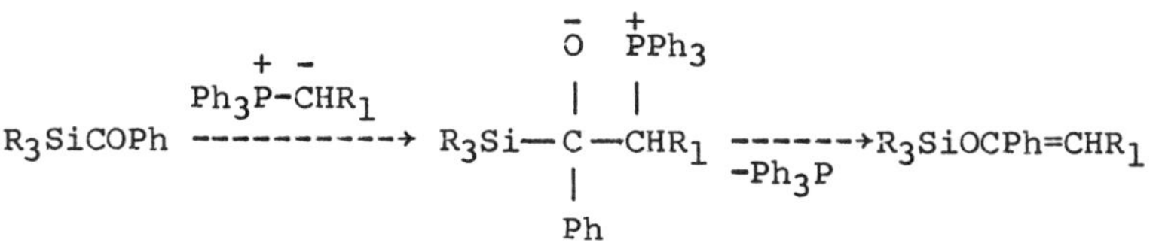

The first of the thermally induced silyl group migrations to be discussed will be the migration in epoxides. In contrast to epoxides decomposing at high temperatures to a variety of products, epoxysilanes

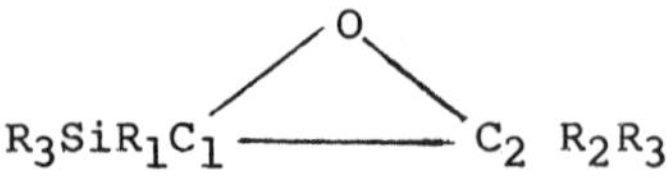

$$R_3SiR_1C_1 \overset{O}{\diagup\!\!\!\diagdown} C_2 R_2R_3$$

rearrange cleanly at 170-310° to siloxyalkenes [134].
This reaction follows two different rearrangement pathways
(Eq. 35), its course being influenced by the substituents
attached to the epoxide ring and their relative geometries.
With some epoxysilanes a "double migration" process (path
a) occurs in which the silyl group migrates from C to oxygen and becomes a siloxy group attached to C, while R_3 or
R_2 migrates from C_2 to C_1. In other cases (path b) the
rearrangement involves only a migration of the silyl group
from C to the oxygen. Each pathway displays high stereospecificity [134] at lower temperatures and may be catalyzed by
impurities [135].

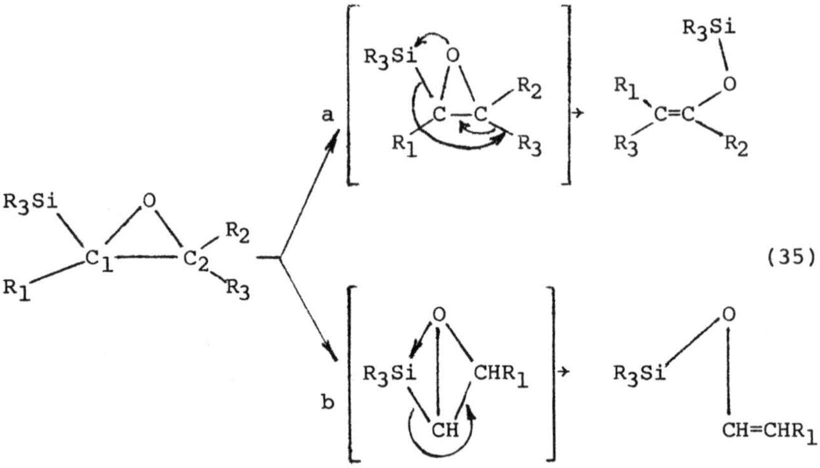

(35)

Silylmethyl allyl ethers undergo rearrangements when
heated to 160-200° and produce silyl ethers [136] (Eq. 36).

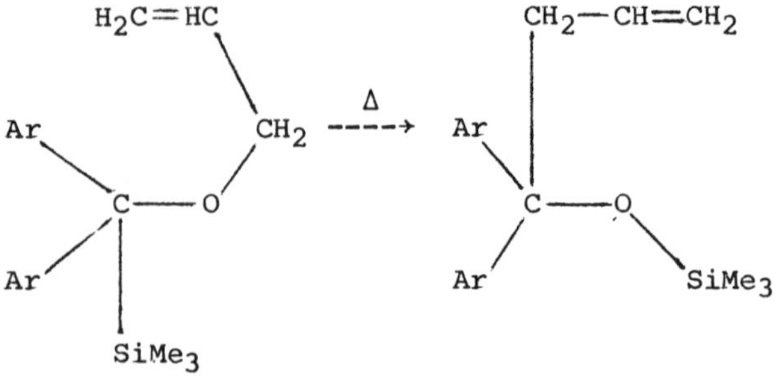

(36)

INTRAMOLECULAR INTERACTION

Despite the absence of appreciable amounts of side products, this reaction does not proceed according to a simple mechanism. The migration of the silyl group is 100% intramolecular, while that of the allyl group is only 72% intramolecular. In the presence of a radical trapping agent the yield decreases, but migration of the allyl group proceeds completely intramolecularly. It appears [136] that the rearrangement of the initial motion of the silyl group is far more advanced than that of the allyl group. The mechanism (Eq. 37) is thus dissimilar to dyotropic rearrangements [124] and consists of the rate-determining formation of a short-lived complex in which the silyl moiety coordinates with the oxygen, followed by rapid decomposition into radicals (path a represents homolysis with neighboring group participation) or an intramolecular concerted rearrangement (path b).

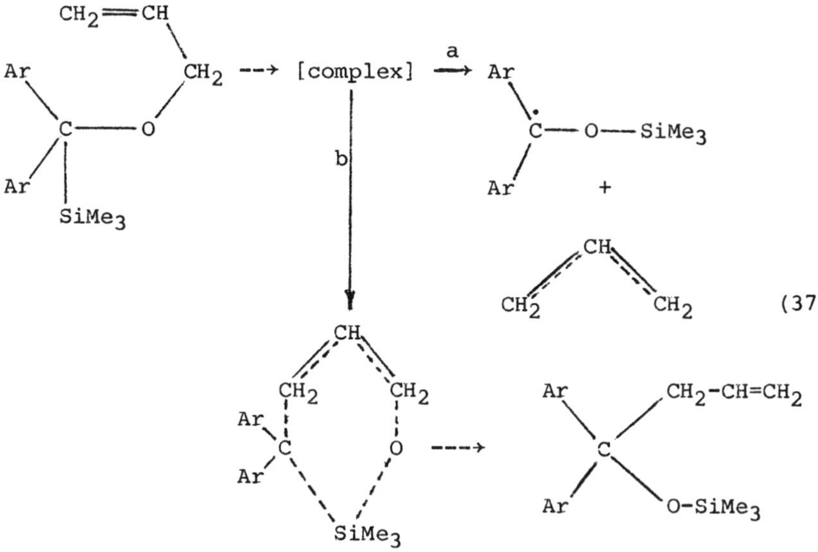

$$\text{(37)}$$

The thermolysis of (methoxymethyl)silanes, a first-order reaction in silane, produces methoxycarbene [137] and may be explained by the transition state in which the cleavage of the Si-C bond is influenced by anchimeric assistance of the methoxy group (Eq. 38).

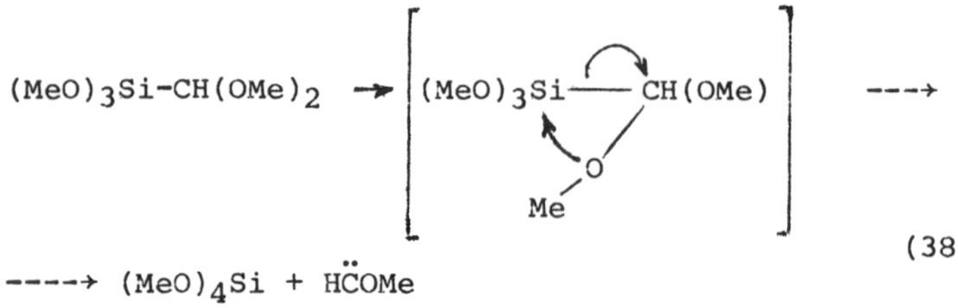

$$(MeO)_3Si-CH(OMe)_2 \rightarrow \left[(MeO)_3Si\underset{\underset{Me}{O}}{\text{—}}CH(OMe) \right] \dashrightarrow$$

$$\dashrightarrow (MeO)_4Si + H\ddot{C}OMe \qquad (38)$$

The silanes

$$\underset{\underset{CH_2ON(CF_3)_2}{|}}{Me_2SiLCHON(CF_3)_2} \qquad (L = Me \text{ or } ON(CF_3)_2)$$

and $Cl_3SiCMe_2ON(CF_3)_2$ rearrange thermally [138] by a process in which the Si-C-O-N system is reorganized into N-C-O-Si (Eq. 39).

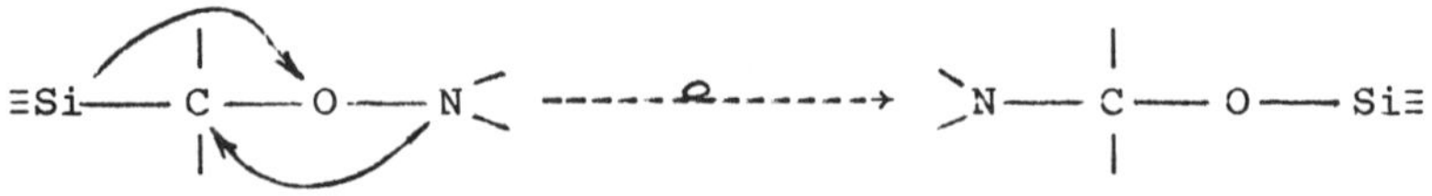

$$\equiv Si\text{—}C\text{—}O\text{—}N{\lt} \dashrightarrow {\gt}N\text{—}C\text{—}O\text{—}Si\equiv \qquad (39)$$

The apparently low ability of the oxygen to coordinate to silicon in these compounds might be increased by the cleavage of the relatively weak N-O bond, a cleavage for which there is precedence. The α-oxygenated benzyltriphenylsilanes undergo [139] thermal rearrangement involving migration of the ester group to silicon and a shift of a phenyl group from silicon to carbon (Eq. 40).

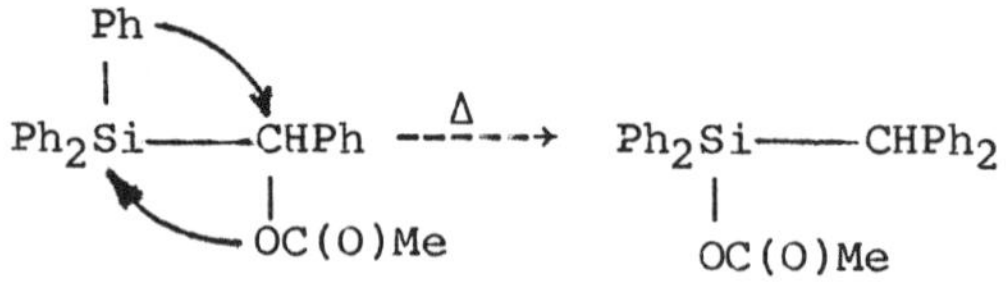

$$\underset{\underset{OC(O)Me}{|}}{Ph_2Si\text{—}CHPh} \xrightarrow{\Delta} \underset{\underset{OC(O)Me}{|}}{Ph_2Si\text{—}CHPh_2} \qquad (40)$$

with Ph migrating.

<u>Interaction With Sulfur</u>. The 1,2-migration of silicon from carbon to sulfur takes place [140] during a thermal or radical-initiated rearrangement of α-trimethylsilylbenzylmercaptan (Eq. 41).

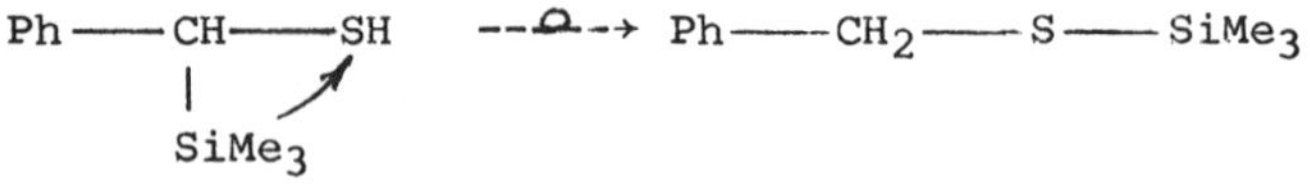

$$\underset{\underset{SiMe_3}{|}}{Ph\text{—}CH\text{—}SH} \xrightarrow{\Delta} Ph\text{—}CH_2\text{—}S\text{—}SiMe_3 \qquad (41)$$

This thermally induced rearrangement requires temperatures

80° lower than those required for the analogous rearrangement of the oxygen compound.

Interaction With Halogens. 1,1,2,2,-Tetrafluoroethylsilanes, $X_nMe_{3-n}SiCF_2CHF_2$, decompose by a first-order homogeneous reaction for which a two-step mechanism was proposed [141,142] and involved (1) the formation of a reactive intermediate by an internal nucleophilic attack on silicon by a fluorine in the α-position, and (2) the subsequent migration of an atom other than fluorine from the β-carbon atom (Eq. 42).

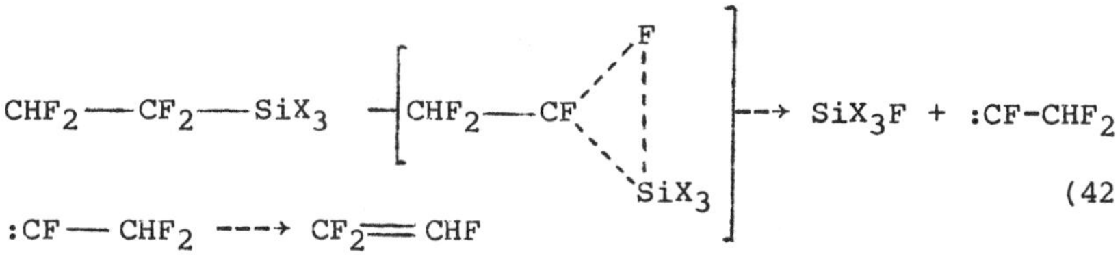

$$CHF_2-CF_2-SiX_3 \longrightarrow \left[CHF_2-CF\begin{smallmatrix}F\\|\\SiX_3\end{smallmatrix} \right] \longrightarrow SiX_3F + :CF-CHF_2$$

$$:CF-CHF_2 \dashrightarrow CF_2=CHF \tag{42}$$

The replacement of the fluorines on the silicon by Me groups leads to a very marked decrease in the rate of decomposition. Keeping in mind the greater electron release by methyl groups to silicon, this decrease can be attributed to the less electropositive silicon and to attack by fluorine on a less receptive silicon in the trimethylsilyl compound.

A similar expulsion of silyl chloride (obviously initiated by coordination of chlorine to silicon) occurs [143] during the reaction of vinylsilanes with palladium salts (Eq. 43).

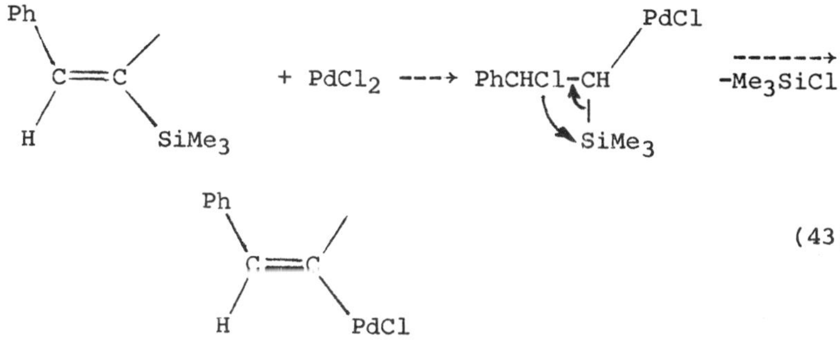

$$\underset{H}{\overset{Ph}{\diagdown}}C=C\underset{SiMe_3}{\diagup} + PdCl_2 \dashrightarrow PhCHCl-\underset{SiMe_3}{\overset{PdCl}{\mid}}CH \xrightarrow{-Me_3SiCl}$$

$$\underset{H}{\overset{Ph}{\diagdown}}C=C\underset{PdCl}{\diagup} \tag{43}$$

Also, α-halogenated alkylsilanes undergo rearrangement involving interchange between the halogen and groups on the silicon. The thermally induced reaction occurs with trichlorosilyl [144] or triphenylsilyl [139] compounds (Eqs. 44 and 45, respectively).

$$Y_2CH\text{---}SiCl_3 \xrightarrow{\Delta} \left[\begin{array}{c} Y \\ | \\ C\text{---}SiCl_2 \\ | \quad \diagup \\ H \quad Cl \end{array} \right] \longrightarrow ClYHC\text{---}SiCl_2Y \quad (44)$$

$$\begin{array}{c} Ph \\ | \\ Ph_2Si\text{---}CHPh \\ | \\ Y \end{array} \xrightarrow{\Delta} \begin{array}{c} \\ Ph_2Si\text{---}CHPh_2 \\ | \\ Y \end{array} \quad (45)$$

Facile alkyl- (or aryl-)chloro group interchange also proceeds [145,146] in the presence of strong electrophilic agents. This reaction is similar to the well-known Wagner-Meerwein rearrangement of carbon chemistry, but it is not complicated by the possibility of migration of alkyl groups from carbon, and the product of migration is not subject to loss of a proton because of the unwillingness of silicon to form a stable double bond. The reaction of aryl(chloromethyl)dimethylsilanes catalyzed with $AlCl_3$ is assumed [146,147] to proceed by a synchronous intramolecular migration of the aryl group and chlorine in the species in which the CH_2Cl group is associated with the catalyst (Eq. 46).

$$ArMe_2SiCH_2Cl \xrightarrow{AlCl_3} \left[\begin{array}{c} Ar \\ \diagup \quad \diagdown \\ Me_2Si\text{---}CH_2 \\ \diagdown \quad \diagup \\ Cl \\ | \\ AlCl_3 \end{array} \right] \longrightarrow ClMe_2SiCH_2Ar \quad (46)$$

The reaction of α-(chloroalkyl)alkyldimethylsilanes catalyzed with SbF_5 in polar nitromethane involves [146] a somewhat earlier dissociation of the C-Cl bond.

Coordination of halogen to silicon can also be presumed to occur during some substitution reactions at the α-carbon. Thus, reaction of H_3SiCH_2Cl with AgF or KF gives H_3SiF and carbene (:CH_2) and was assumed [148] to proceed as written in Eq. (47),

$$H_3Si\text{---}CH_2Cl \xrightarrow{KF} \left[H_3Si\text{---}CH_2\text{---}F \right] \longrightarrow H_3SiF + :CH_2 \quad (47)$$

and preparation of Me_3SiCH_2F by reaction of Me_3SiCH_2OL (L = H, p-$MeC_6H_4SO_2$) with Yarovenko reagent, Et_2NCF_2CHFCl, is complicated [99] by the simultaneous formation of Me_2EtSiF. Similarly, the F^- rearrangement with prior attack at silicon has been suggested [149] as an explanation for conversion of Me_3SiCH_2I to Me_2EtSiF, and a similar mechanism can explain fluoride ion-induced ring enlargement of tricyclic silicon heterocycles [150].

2.3.2 Reactions Involving 1,3-Interaction

<u>Interaction With Carbon</u>. A 1,3-migration of silicon from sp^3 to sp^2 or sp carbon occurs in allyl- and propargylsilanes and is induced by heat [151].

$$RMe_2Si\text{---}CR_2\text{---}CR_3 \xrightarrow{\Delta} RMe_2Si\text{---}CR_2\text{---}CR_3 \quad (48)$$

The first reaction (Eq. 48) follows a unimolecular rate law up to 90% completion and is accompanied by inversion of the silicon configuration. Its negative activation entropy suggests a cyclic concerted transition state.

The second reaction (Eq. 49) is reversible and is in all respects similar to the silaallylic rearrangement.

$$Me_3SiCH_2\text{---}C\equiv CH \underset{\Delta}{\rightleftharpoons} CH_2=C=CHSiMe_3 \quad (49)$$

Under similar conditions in the carbon analogue, when fragmentation reactions become evident, neither rearrangement occurs, which is an indication [151] that the decisive factor may be the dissociation energy of the critical bond, i.e., the bond between the allylic (or propargylic) framework and the (potential) migrating center in the substrate. The normal carbon-silicon bond dissociation energy (ca. 290 kJ/mol) is evidently below the threshold value necessary to foster the rearrangement reaction in competition with the homolytic fragmentation mode. The normal carbon-carbon bond energies (ca. 330-360 kJ/mol) are considerably (ca. 42 kJ/mol) above this threshold.

A 1,3-migration of the silyl group in 1-silyl-3-trimethylsilyl-3-phenylpropynes proceeds [152] upon irradiaton with UV light and is reversible (Eq. 50). The equilibrium favors the formation of silylphenylpropadienes.

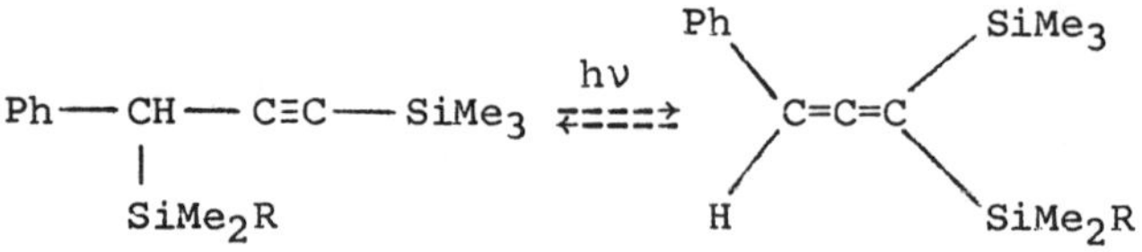

$$Ph-CH(SiMe_2R)-C\equiv C-SiMe_3 \underset{h\nu}{\rightleftarrows} (Ph)(H)C=C=C(SiMe_3)(SiMe_2R) \qquad (50)$$

Interaction With Nitrogen. The β-(aminoethyl)silanes, e.g., $Me_3Si(CH_2)_2NHR$, decompose in the presence of n-BuLi into a variety of products by a reaction that possibly includes a four-center transition state in which anionic nitrogen intramolecularly coordinates with silicon [125] (Eq. 51).

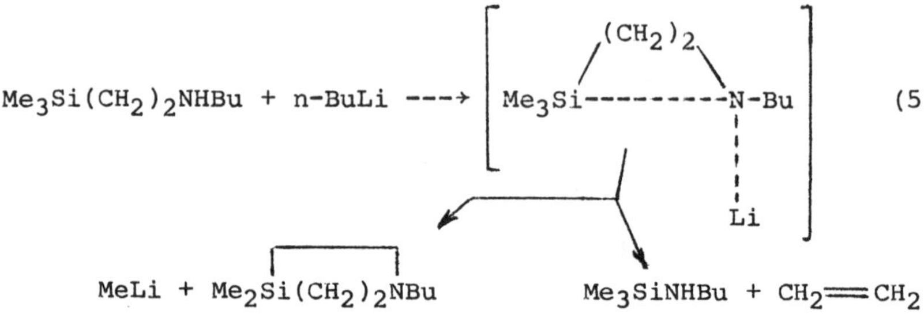

$$Me_3Si(CH_2)_2NHBu + n\text{-BuLi} \longrightarrow \left[Me_3Si\cdots(CH_2)_2\cdots N(Bu)\cdots Li \right] \qquad (51)$$

$$\swarrow \qquad \searrow$$

$$MeLi + Me_2Si(CH_2)_2NBu \qquad Me_3SiNHBu + CH_2=CH_2$$

It is probable that an intramolecular 1,3 rearrangement of the silyl group to nitrogen also occurs during the isomerization of silylated aldimines to N-silyl-N-alkylenamines (Eq. 52). The rate of this reaction decreases with the increasing volume of the substitutents attached to the nitrogen [153].

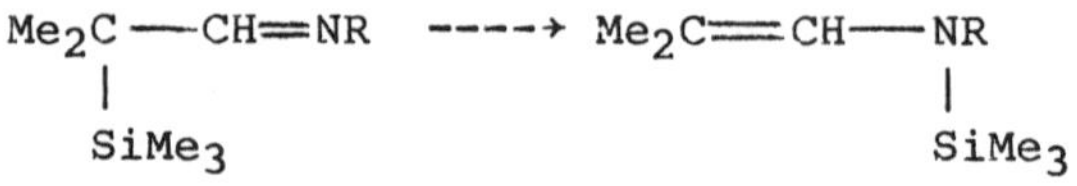

$$Me_2C(SiMe_3)-CH=NR \longrightarrow Me_2C=CH-NR(SiMe_3) \qquad (52)$$

Interaction With Oxygen. The β-ketosilane to

INTRAMOLECULAR INTERACTION

siloxyalkene thermal rearrangement [129,154] follows first-order kinetics, and its energy of activation and negative $\Delta S^{\ddagger}$ values are unaltered with large changes in structure of the reagents. The reaction (Eq. 53) proceeds with retention of configuration at silicon.

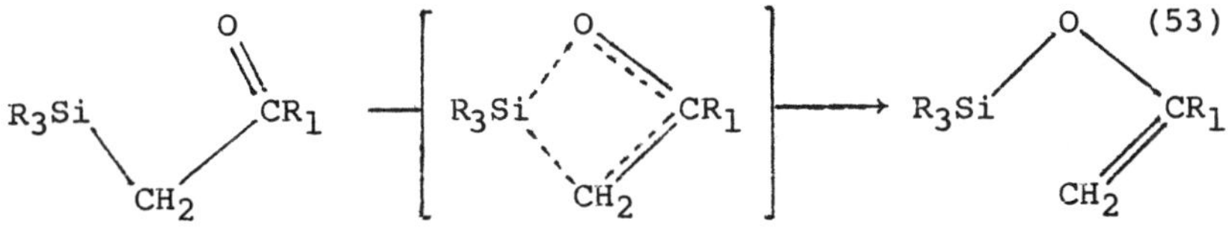

$$R_3Si-CH_2-C(=O)-CR_1 \longrightarrow [R_3Si\cdots O\cdots CR_1\cdots CH_2] \longrightarrow R_3Si-O-CR_1=CH_2 \quad (53)$$

The data are consistent with the rate-determining formation of a five-coordinate silicon intermediate, wherein the Si-O bond formation is the deriving force of the reaction [155-156]. A similar mechanism, i.e., an intramolecular process involving a four-center transition state, is also assumed for the base-catalyzed or thermal rearrangement of silylphenylacetic acid [157,158] (Eq. 54),

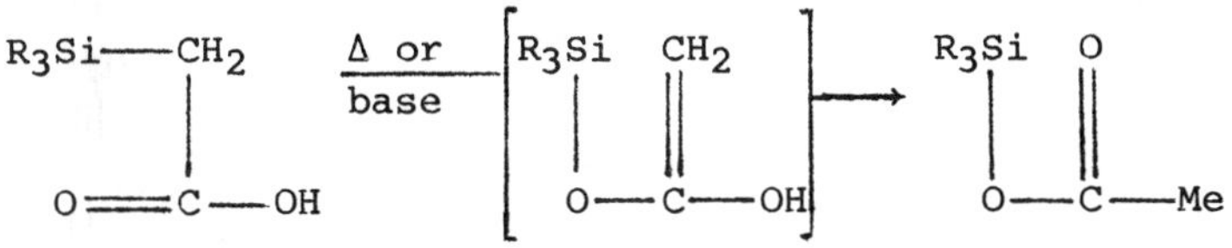

$$\underset{O=C-OH}{R_3Si-CH_2} \xrightarrow[\text{base}]{\Delta \text{ or}} \left[\underset{O-C-OH}{R_3Si\ CH_2}\right] \longrightarrow \underset{O-C-Me}{R_3Si\ O} \quad (54)$$

and the thermal rearrangement of β-silyl sulfoxides [159,160] (Eq. 55).

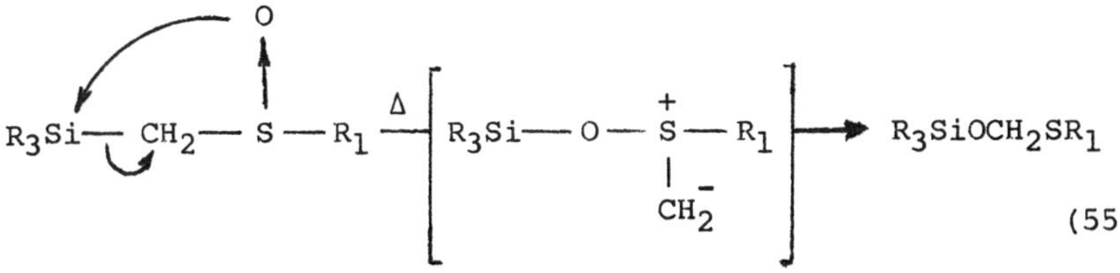

$$R_3Si-CH_2-S(=O)-R_1 \xrightarrow{\Delta} \left[R_3Si-O-\overset{+}{S}(CH_2^-)-R_1\right] \longrightarrow R_3SiOCH_2SR_1 \quad (55)$$

The ease of 1,3 migration of silicon in other similar systems is influenced by substituents. The rearrangement of R_3SiCH_2COMe is hindered by progressive attachment of chlorines in place of alkyl gorups [161]. The -I effect of chlorines increases the electrophilic ability of silicon but simultaneously decreases the nucleophilic ability of the carbonyl oxygen; the latter effect obviously has a great impact on the ease of the formation of the O→Si coordination bond.

The rearrangement in trialkylsilylketene acetals is

facilitated by electron-withdrawing substituents on nitrogen [162] (Eq. 56).

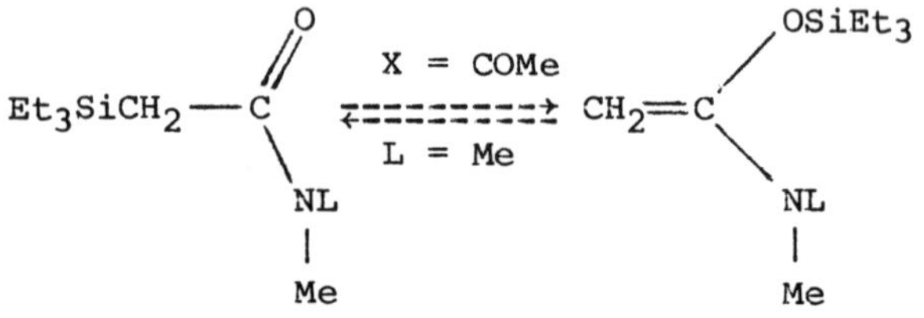

$$Et_3SiCH_2-\underset{\underset{Me}{\overset{|}{NL}}}{\overset{\overset{O}{\parallel}}{C}} \underset{L = Me}{\overset{X = COMe}{\rightleftarrows}} CH_2=\underset{\underset{Me}{\overset{|}{NL}}}{\overset{OSiEt_3}{C}} \qquad (56)$$

The course of numerous reactions can be understood in terms of an initially formed O→Si coordination bond followed by either a 1,3 silyl migration from carbon to oxygen or by elimination of R_3SiOL molecules. Included are, e.g., thermal rearrangement of tetramethylene-[(α-trimethylsilyl)-phenacyl]sulfonium chloride [163], decomposition of silyl phosphonates [164], reaction of silylketene with phosphonates or thiophosphonates [165], thermal decomposition of cis-2-trimethylsilylcyclohexyl methyl ether and the acyclic $Me_3Si(CH_2)_2OMe$ ether [166], reaction of trimethylsilylmethylphosphonium salts with ketones [167] (Eqs. 57-61),

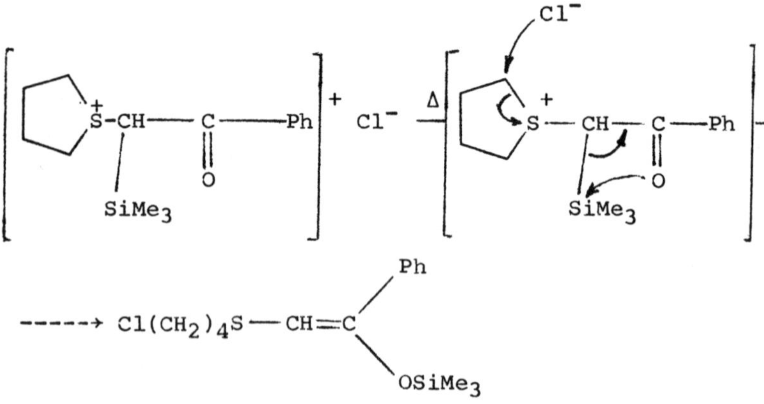

$$\left[\underset{\underset{SiMe_3}{\overset{|}{}}}{\overset{+}{S}}-CH-\underset{\overset{\parallel}{O}}{C}-Ph\right]^+ Cl^- \xrightarrow{\Delta} \left[\overset{+}{S}-CH-\underset{\overset{\parallel}{O}}{C}-Ph\right] \longrightarrow \qquad (57)$$

$$\dashrightarrow Cl(CH_2)_4S-CH=C\underset{OSiMe_3}{\overset{Ph}{\diagup}}$$

$$\underset{CH_2-P(O)OR}{\overset{Me_3Si \quad OR}{|}} \dashrightarrow Me_3SiOr + CH_2=P(O)OR \qquad (58)$$

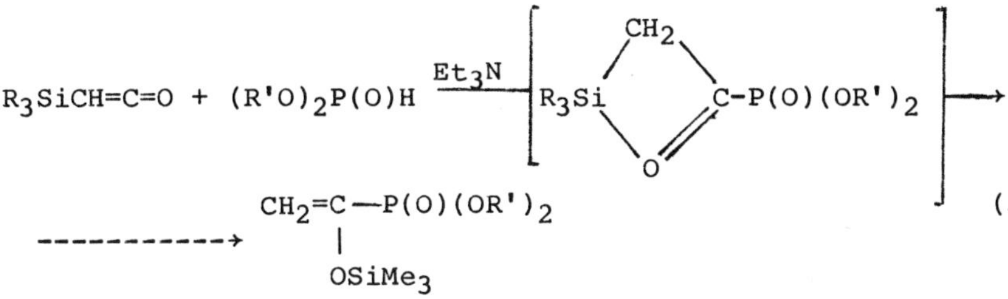

INTRAMOLECULAR INTERACTION

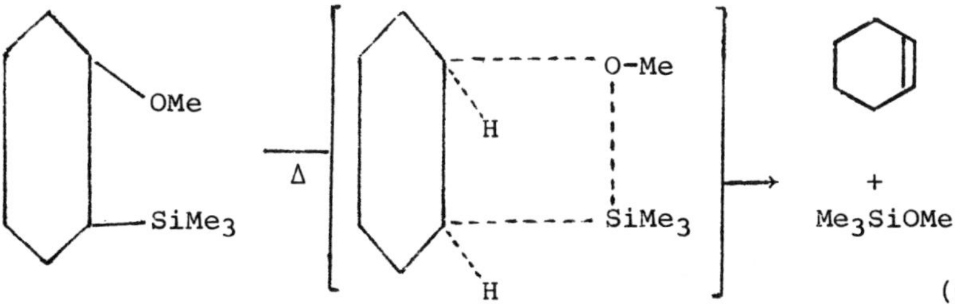

$$R_3SiCH=C=O + (R'O)_2P(O)H \xrightarrow{Et_3N} \left[R_3Si \underset{O}{\overset{CH_2}{\diamond}} C-P(O)(OR')_2 \right] \longrightarrow$$

$$\dashrightarrow \underset{OSiMe_3}{CH_2=C-P(O)(OR')_2} \tag{59}$$

$$\text{[cyclohexene with OMe and SiMe}_3\text{]} \xrightarrow{\Delta} \left[\text{transition state with O-Me, SiMe}_3, H, H} \right] \longrightarrow \text{cyclohexene} + Me_3SiOMe \tag{60}$$

$$Me_3Si\overset{-}{C}H-\overset{+}{P}Ph_3 + Ph_2C=O \longrightarrow \left[\begin{array}{c} Me_3Si-\overset{+}{C}HPPh_3 \\ \updownarrow \\ O \longrightarrow CPh_2 \end{array} \right] \longrightarrow \begin{array}{c} Me_3SiO^- \\ + \\ Ph_2C=CHPPh_3 \end{array} \tag{61}$$

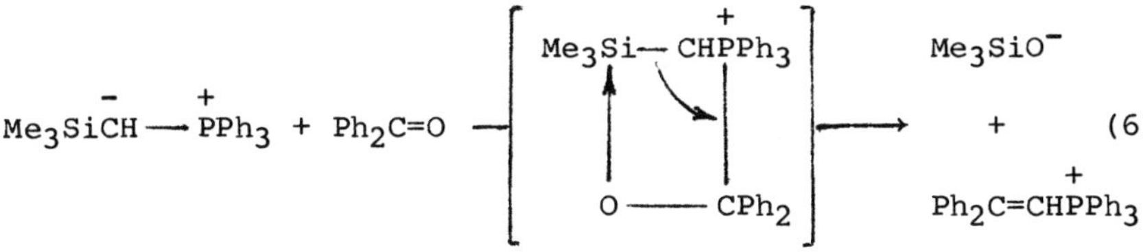

formation of Me_3SiOEt from $(CO)_5CrC(OEt)CH_2SiMe_3$ in the mass spectrometer [168], and a whole family of reactions between α-silylsubstituted organometallic compounds of general formula $R_3Si-CR_1R_2-M$ with aldehydes or ketones (Peterson reaction, [169]) that proceed through an intramolecular attack of the silicon by oxygen in a four-center process and thus provide the basis for a variety [170-175] of useful olefin syntheses (Eq. 62).

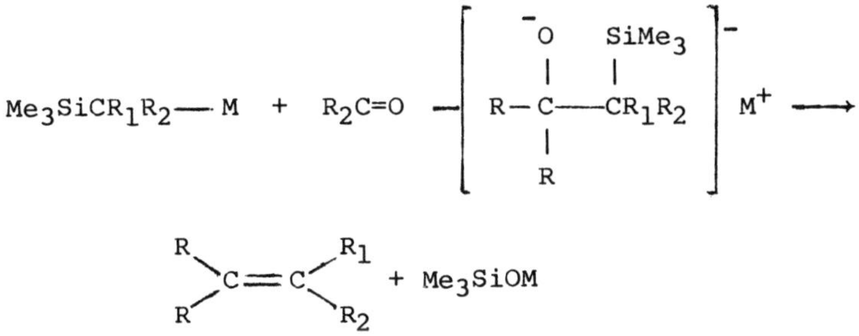

$$Me_3SiCR_1R_2-M + R_2C=O \longrightarrow \left[\begin{array}{c} ^-O \quad SiMe_3 \\ | \quad\quad | \\ R-C-CR_1R_2 \\ | \\ R \end{array} \right] M^+ \longrightarrow \tag{62}$$

$$\underset{R}{\overset{R}{>}}C=C\underset{R_2}{\overset{R_1}{<}} + Me_3SiOM$$

Interaction With Halogens. The gas-phase [176-179] and liquid phase [180] decomposition of 2,2-difluoroethylsilanes, $X_nR_{3-n}SiCH_2CHF_2$, are first-order homogeneous reactions that are insensitive to the presence of radical inhibitors and are presumed to occur by a unimolecular transfer of β-fluorine to silicon in a four-center transition state (Eq. 63).

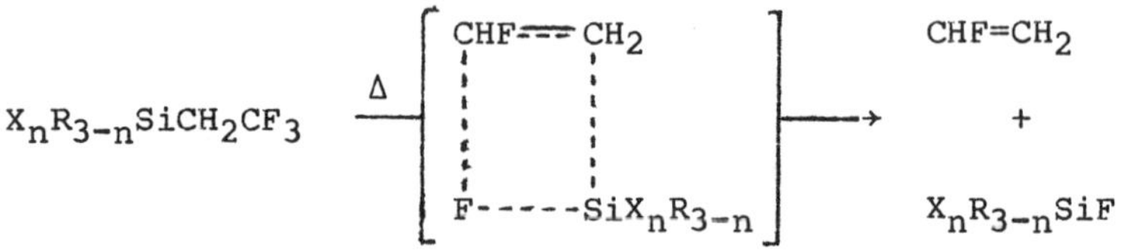

$$X_nR_{3-n}SiCH_2CF_3 \xrightarrow{\Delta} \left[\begin{array}{c} CHF = CH_2 \\ | \quad \quad | \\ F \text{-----} SiX_nR_{3-n} \end{array} \right] \longrightarrow \begin{array}{c} CHF=CH_2 \\ + \\ X_nR_{3-n}SiF \end{array} \quad (63)$$

The reaction is faster in solution than in the gas phase, thus indicating that the transition state is more heavily solvated than the reactants and therefore is more polar in character than the reactant molecules.

A similar mechanism is also operative in the pyrolysis of 2,2-(difluoroethyl)polysiloxane [181] (Eq. 64), the gas-phase decomposition of 2-(chloroethyl)silanes $X_nR_{3-n}Si(CH_2)_2Cl$ [182-184], and probably in the decomposition of 2-chloro-4-silylpropyldichlorophosphine oxide [185] (Eq. 65).

$$\begin{array}{c} CHF \text{---} CH_2 \\ | \quad \quad | \\ F \quad \quad SiO_{1.5} \end{array} \longrightarrow CHF=CH_2 + FSiO_{1.5} \quad (64)$$

$$Me_3Si\text{---}CH_2\text{---}CH\text{---}CH_2P(O)Cl_2 \xrightarrow[-MeSiCl]{} CH_2=CH\text{---}CH_2\text{---}P(O)Cl_2 \quad (65)$$
$$\quad \quad \quad \quad \quad | \quad$$
$$\quad \quad \quad \quad \quad Cl$$

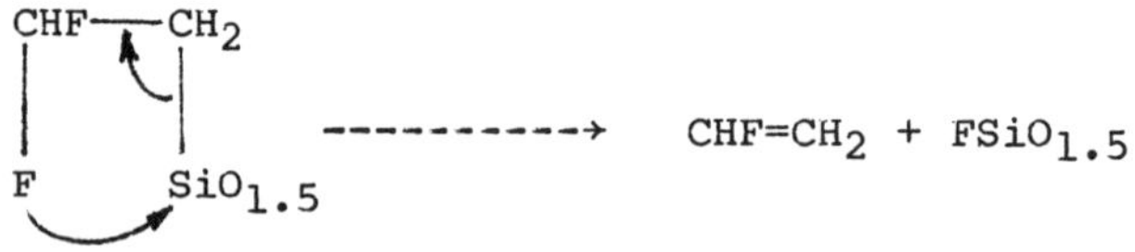

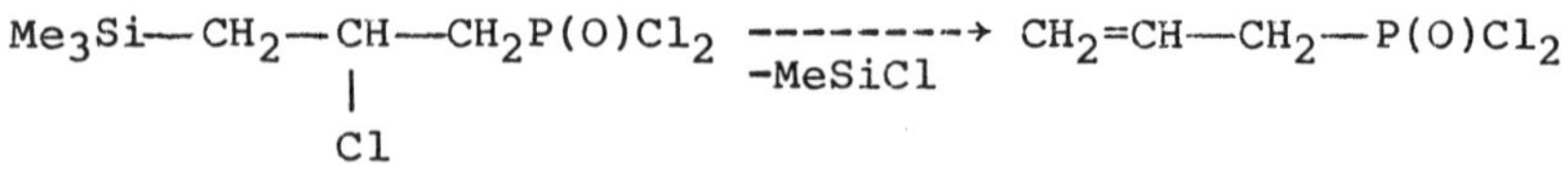

2.3.3 Reactions Involving 1,4-Interaction

Interaction With Carbon. A 1,4-interaction between silicon and carbon takes place primarily when carbon possesses electron density sufficient to enable nucleophilic attack on silicon. Thus, 1,4-anionic rearrangements

INTRAMOLECULAR INTERACTION

of triorganosilyl groups that are initiated by nucleophilic attack of anionic carbon are known to proceed via 2-triorganosilylethylammonium ion yield intermediates [186] in reactions of N-(triorganosilylmethyl)-[187] and N-(2-triorganosilylethyl)dialkylamines [188] with benzyne, and in ring-closure reactions of 4-lithiobutylsilanes [189,190] (Eqs. 66-69, respectively). Intramolecular cyclization of dimethyl(1-pent-4-enyl)silyllithium cannot be placed into this category of reaction, since it is the silicon in this compound that is electron rich and can serve as a nucleophilic center. The transition state of this reaction, which proceeds with greater facility than the reaction of the analogous 1-hex-5-enyllithium derivative [191], is shown in (Eq. 70).

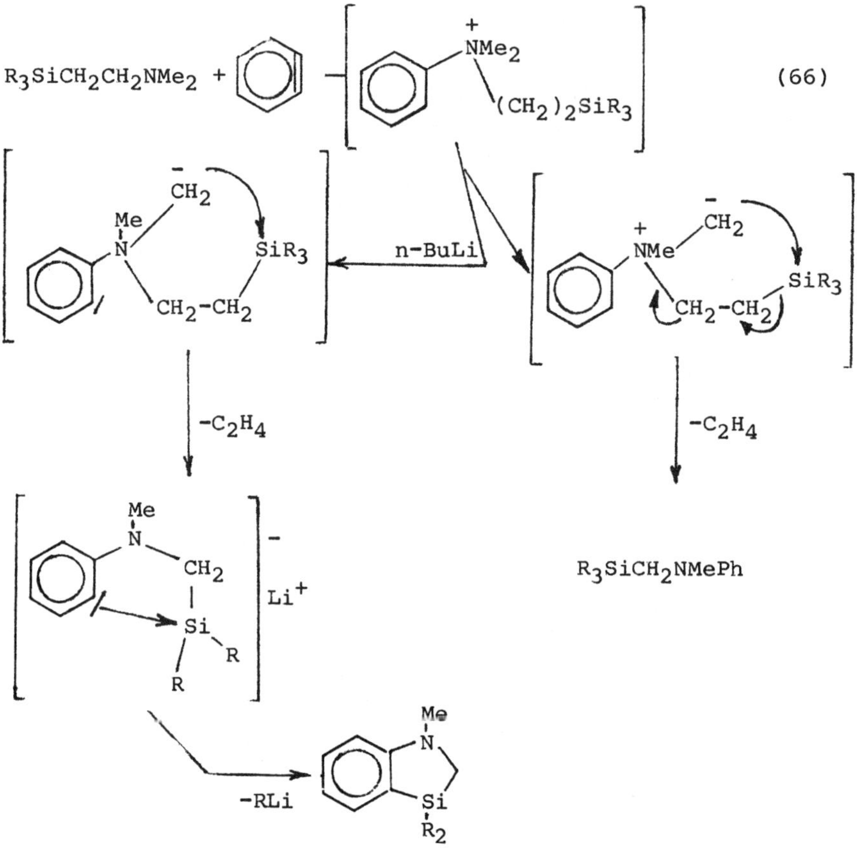

$$R_3SiCH_2CH_2NMe_2 + \text{[benzene]} \rightarrow \left[\text{[Ar-NMe}_2^+ \text{(CH}_2)_2SiR_3] \right] \quad (66)$$

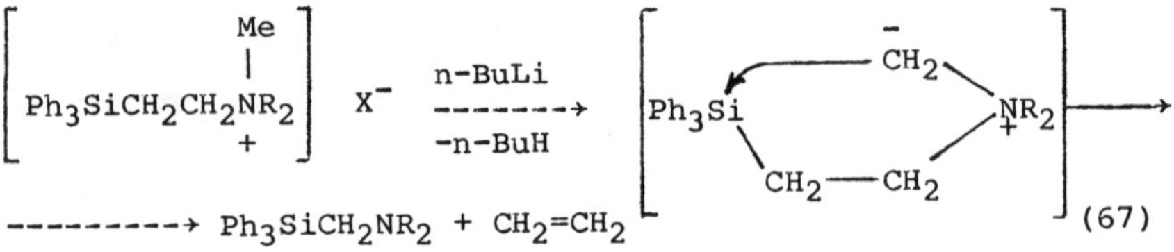

$$\left[\begin{array}{c}\text{Me}\\|\\\text{Ph}_3\text{SiCH}_2\text{CH}_2\text{NR}_2\\+\end{array}\right]X^- \xrightarrow[-n\text{-BuH}]{n\text{-BuLi}} \left[\begin{array}{c}\text{Ph}_3\text{Si}\overset{\curvearrowleft}{\underset{\text{CH}_2-\text{CH}_2}{\overset{\text{CH}_2}{\diagup}}}\overset{-}{\underset{+}{\text{NR}_2}}\end{array}\right] \longrightarrow$$

$$\longrightarrow \text{Ph}_3\text{SiCH}_2\text{NR}_2 + \text{CH}_2=\text{CH}_2 \quad (67)$$

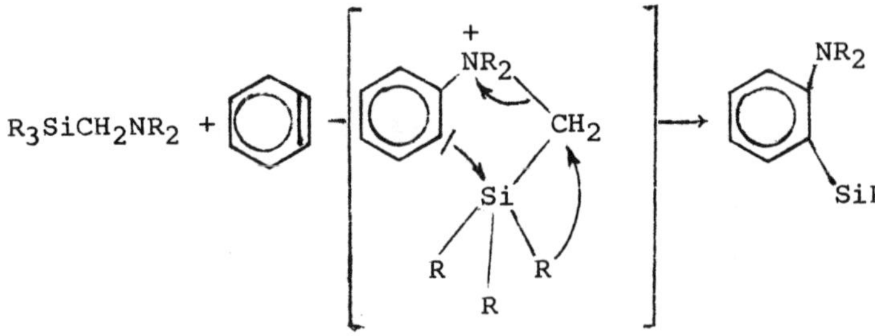

(68)

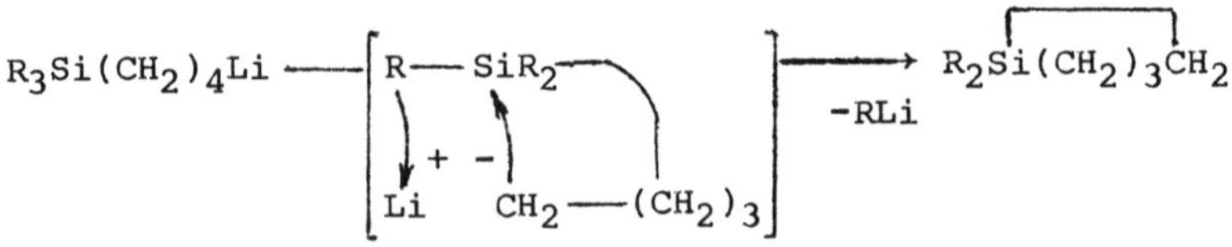

(69)

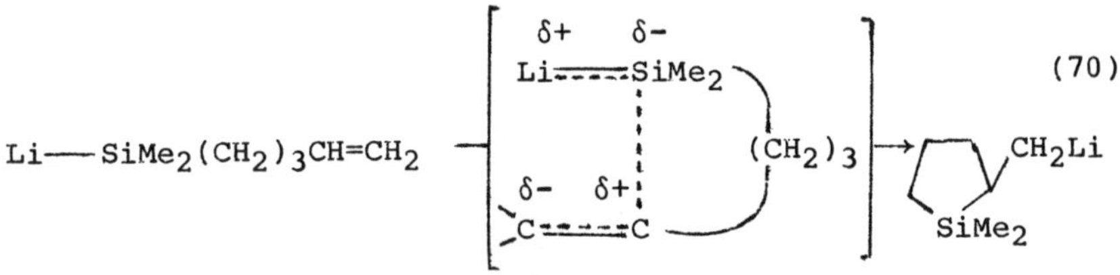

(70)

Interaction With Nitrogen. A 1,4-migration of a silyl group from carbon to nitrogen takes place upon heating of silylated amidines [192] (Eq. 71).

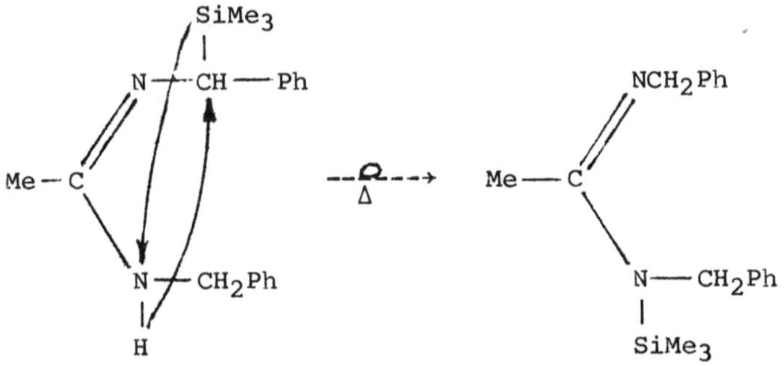

(71)

INTRAMOLECULAR INTERACTION

Also, 3-ethoxysilylpropylaminosilanes undergo a reaction upon heating that can be classified as an intramolecular disproportionation initiated by nucleophilic attack of nitrogen at silicon [193] (Eq. 72).

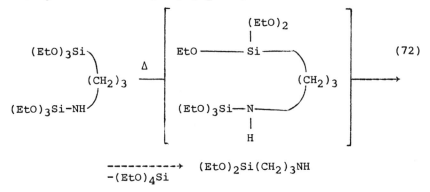

$$\xrightarrow[-(EtO)_4Si]{} (EtO)_2Si(CH_2)_3NH$$

Interaction With Oxygen. A 1,4-interaction between silicon and oxygen occurs primarily in the transition state of cyclization reactions; the nucleophilic attack of oxygen at silicon may be viewed as initiating an intramolecular coupling of substituents (X_1, X_2) bonded to both interacting centers (Eq. 73).

$$\begin{matrix} X_1 - Si \\ \downarrow \quad \uparrow \quad (-C-)_3 \\ X_2 - O \end{matrix} \dashrightarrow \begin{matrix} X_1 \\ | \\ X_2 \end{matrix} + \begin{matrix} Si \\ | \quad (-C-)_3 \\ O \end{matrix} \qquad (73)$$

These intramolecular disproportionations occur during dehydration of 3-hydroxypropyl[bis(trimethylsiloxy)]-methylsilane [194] (Eq. 74),

$$\begin{matrix} Me_3SiO \quad Me \\ \quad \searrow \swarrow \\ \quad Si \\ \quad \diagup \quad \diagdown \\ Me_3SiO \quad (CH_2)_3 \\ \quad \diagdown \\ \quad HO \end{matrix} \xrightarrow{\Delta} Me_3SiOH + (Me_3SiO)MeSi(CH_2)_3O \qquad (74)$$

in thermal cyclization of n-butoxysilyl substituted propoxytrimethylsilanes [195] (Eq. 75),

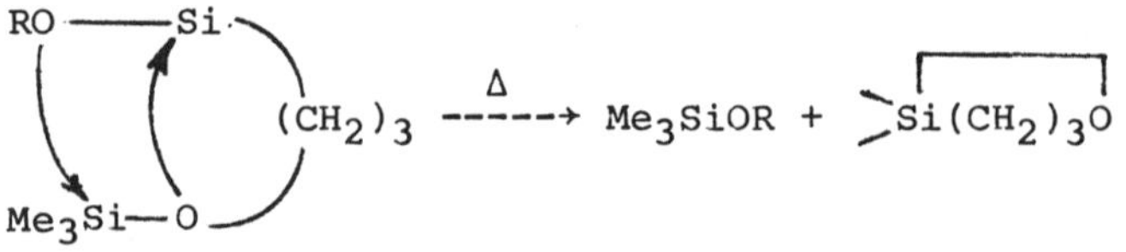

$$\text{RO—Si} \overset{\curvearrowright}{\underset{Me_3Si—O}{\bigg(}} (CH_2)_3 \xrightarrow{\Delta} Me_3SiOR + \overline{{>}Si(CH_2)_3O} \qquad (75)$$

in dehydrocondensation of 3-hydroxypropylsilicon hydrides [196,197] (Eq. 76),

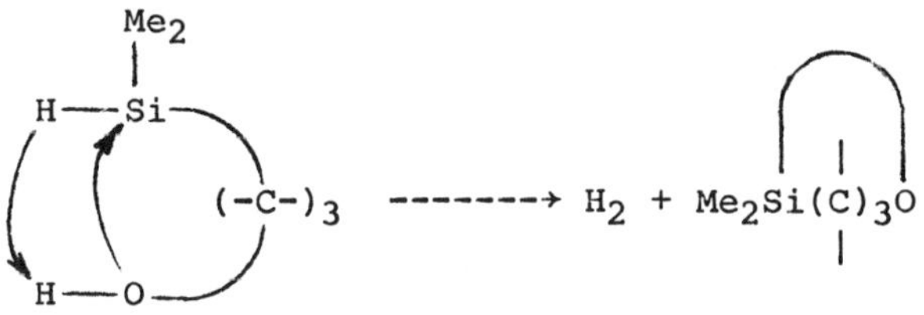

$$\underset{H—O}{\overset{Me_2}{\underset{H—Si}{\bigg(}}} (-\overset{|}{C}-)_3 \xrightarrow{} H_2 + Me_2Si(\overset{|}{\underset{|}{C}})_3O \qquad (76)$$

in dehydration of 1,3-bis(3-hydroxypropyl)tetramethyl-disiloxane [198] (Eq. 77),

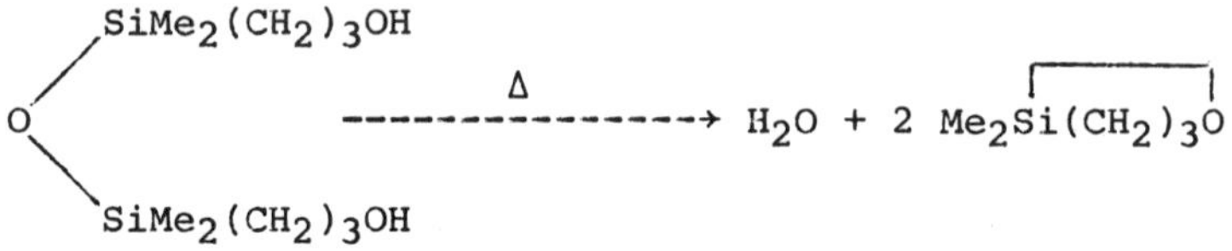

$$O{\diagdown\atop\diagup}{SiMe_2(CH_2)_3OH \atop SiMe_2(CH_2)_3OH} \xrightarrow{\Delta} H_2O + 2\ \overline{Me_2Si(CH_2)_3O} \qquad (77)$$

and in dehydration of 4,4,6,6-tetramethyl-5-oxa-4,6-disilanonane-1,9-dioic acid [199] (Eq. 78).

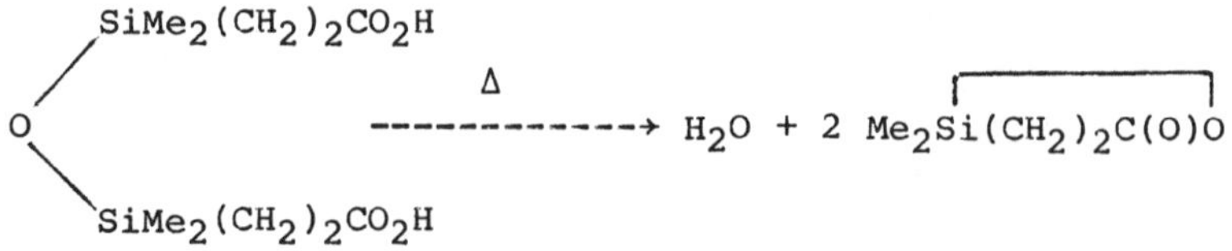

$$O{\diagdown\atop\diagup}{SiMe_2(CH_2)_2CO_2H \atop SiMe_2(CH_2)_2CO_2H} \xrightarrow{\Delta} H_2O + 2\ \overline{Me_2Si(CH_2)_2C(O)O} \qquad (78)$$

The base-catalyzed cyclization [200] of (3-ethoxysilylpropoxy)trimethylsilanes is obviously induced by the interaction of the organosilicon compound with the active part of the catalyst. The formation of the four-center transition state (Eq. 79)

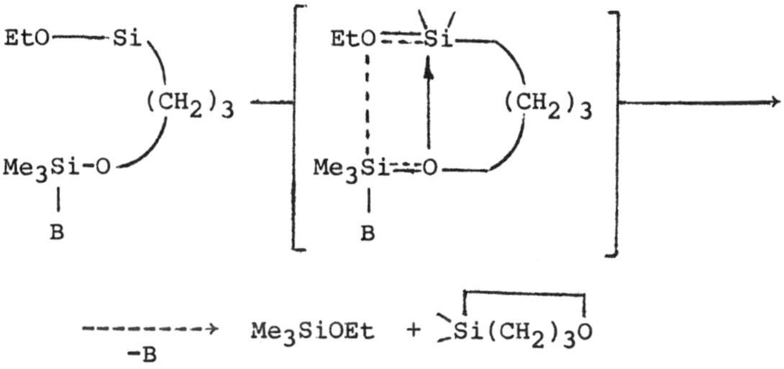

$$\begin{array}{c} \text{EtO—Si} \\ \phantom{\text{EtO—Si}}(CH_2)_3 \\ \text{Me}_3\text{Si-O} \\ | \\ B \end{array} \longrightarrow \left[\begin{array}{c} \text{EtO}\!\cdots\!\text{Si} \\ \vdots\uparrow(CH_2)_3 \\ \text{Me}_3\text{Si}\!\cdots\!\text{O} \\ | \\ B \end{array} \right] \longrightarrow \quad (79)$$

$$\xrightarrow[-B]{} \text{Me}_3\text{SiOEt} + \overline{{>}\text{Si}(CH_2)_3 O}$$

can be assumed to be initiated by nucleophilic attack by the base on the silicon of the trimethylsiloxy group. This attack would increase the electron density at the oxygen, and thus its electron-donor ability, to such an extent that the oxygen may form a coordination bond with the silicon of the ethoxysilyl group. The attack will induce a twofold synchronous nucleophilic substitution at both silicon atoms that would be completed by the release of the active part of the catalyst. The reactivity of the silanes increases with an increasing number of ethoxy groups, which can be explained [200] in terms of the progressive increase of electrophilic character of silicon in the $(\text{EtO})_n\text{Me}_{3-n}\text{Si}$ moiety due to an increasing influence of the -I effect of the EtO group.

The nucleophilic attack of oxygen on silicon also determines the course of the sulfonation of 1-sila-3-cyclopentenes and the corresponding silabicyclohexanes [201] (Eq. 80 and 81).

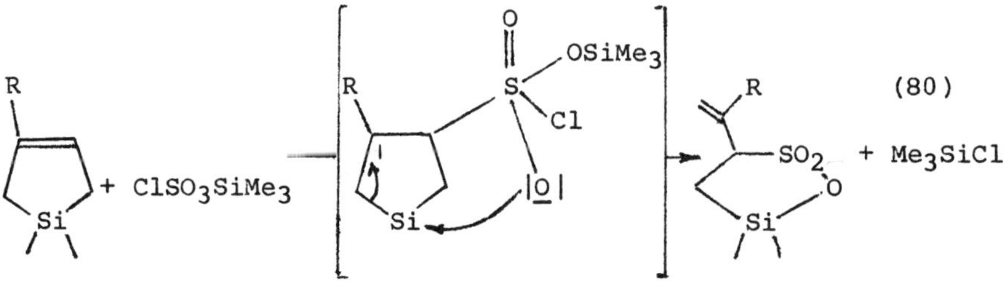

(80)

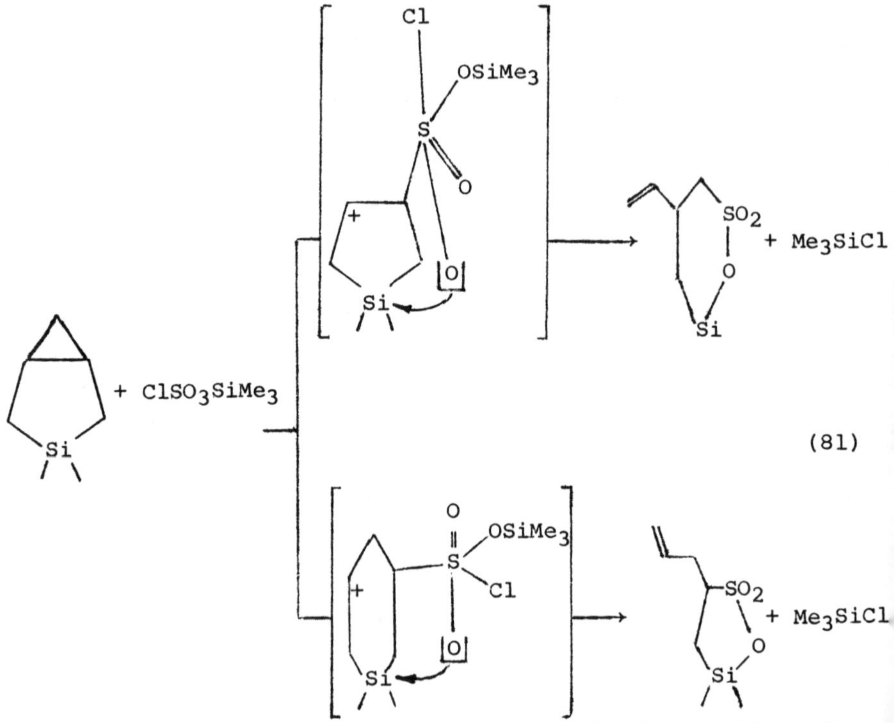

(81)

The same attack is also important in the reaction of α,N-diarylnitrones with lithio trimethylsilyl compounds [202] (Eq. 82),

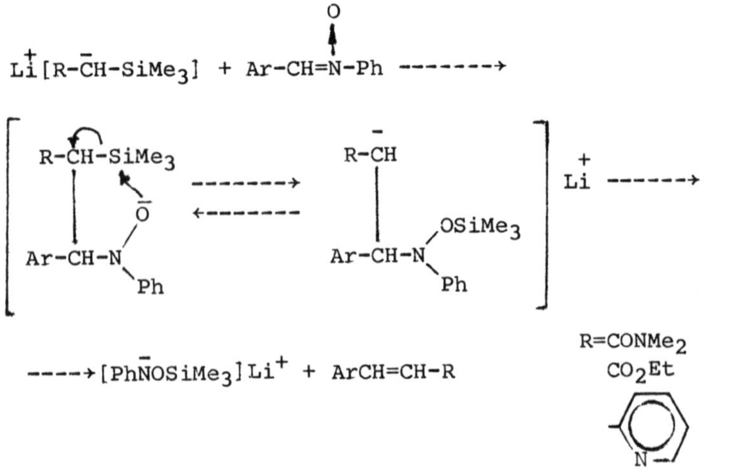

$$\text{Li}^+[\text{R-}\overline{\text{CH}}\text{-SiMe}_3] + \text{Ar-CH=N-Ph} \xrightarrow{\text{O}\uparrow} $$

$$\left[\begin{array}{c} \text{R-CH-SiMe}_3 \\ | \\ \overline{\text{O}} \\ | \\ \text{Ar-CH-N} \\ \quad\quad\text{Ph} \end{array} \xdashrightarrow{\quad} \begin{array}{c} \text{R-}\overline{\text{CH}} \\ \\ \text{Ar-CH-N}^{\text{OSiMe}_3} \\ \quad\quad\text{Ph} \end{array} \right] \text{Li}^+ \longrightarrow \quad (82)$$

$$\longrightarrow [\text{Ph}\overline{\text{N}}\text{OSiMe}_3]\text{Li}^+ + \text{ArCH=CH-R}$$

R=CONMe$_2$
CO$_2$Et
(2-pyridyl)

INTRAMOLECULAR INTERACTION

and destabilizes N-oxides [203] (Eq. 83).

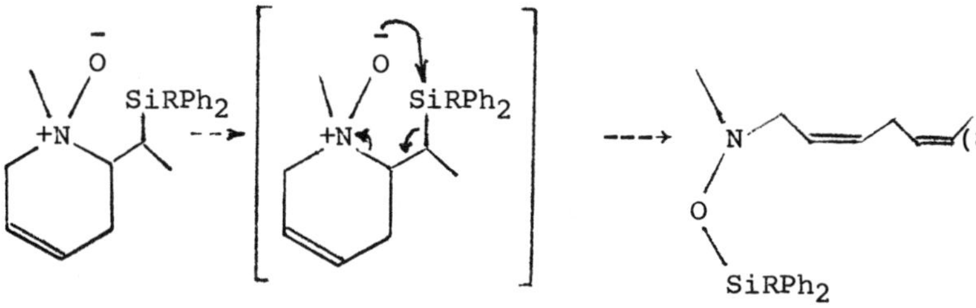

Trans-silylation of bis(trimethylsilyl)acetamide with ClCH$_2$SiMe$_2$Cl gives [204] (O-Si)-chloro[(N-chlorodimethyl-silylacetamido)methyl]dimethylsilane, a compound that not only is unusual because it was formed by substitution at the ClCH$_2$ group rather than at the silicon atom, but also by intramolecular O→Si dative bonding in the ground state (Eq. 84).

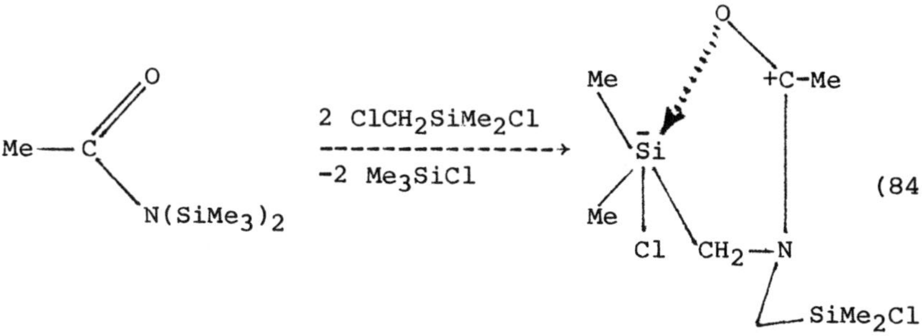

Strong coordination of the oxygen to silicon within a 5-membered ring also occurs [205] in amides, RCONR'[CH$_2$SiMe$_2$Cl], which exist almost solely as the rotamer with the ClMe$_2$Si group cis to the carbonyl:

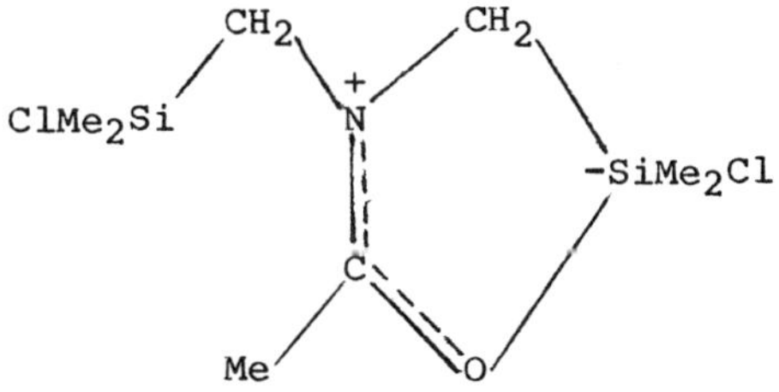

All of these compounds serve as stable representatives of the five coordinate type of structure that has been postulated as being involved in a variety of reactions of carbon-functional silanes.

The mass spectra of carbon-functional silanes also show remarkable differences from the analogous organic molecules in which silicon is replaced by carbon. The different behavior of silicon compounds under electron impact is due to the strong interaction of a distant electron-rich carbon-functional group either with a neutral quaternary silyl center or with a siliconium ion center formed from the former by loss of a Me group. The first type of interaction, which is relevant to the topic of this chapter, involves the direct transfer of an intact Me$_3$Si group from one part of the ion to another (here to the oxygen with which it becomes directly bonded) with a concurrent fragmentation in a manner similar to certain types of specific hydrogen migrations. This type of behavior was observed with 4,4-dimethylsilacycloheptanone [206] (Eq. 85).

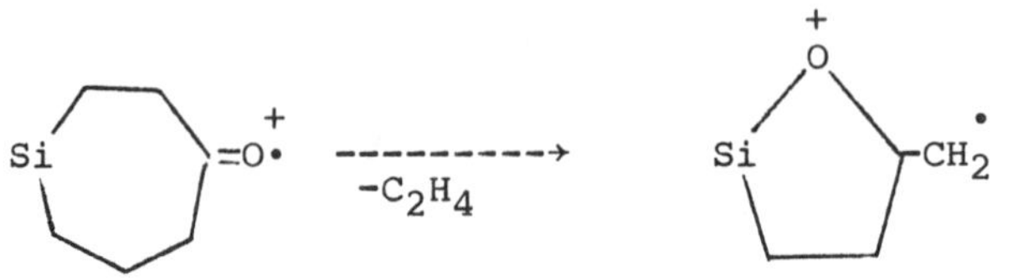

$$\text{(Eq. 85)}$$

Interaction With Halogens. The 3-halogenopropylsilane compounds are more stable than their 2- or 1-alkylsubstituted homologues, and they decompose appreciably only at temperatures of 350-400°. The mechanism of the decomposition is dependent on the substituents at silicon, and it involves heterolytic Si-C bond fission that is assumed [207] to be initiated by a nucleophilic attack of fluorine on silicon in the $(CF_3CH_2CH_2)_nSiR_{4-n}$ compounds (Eq. 86).

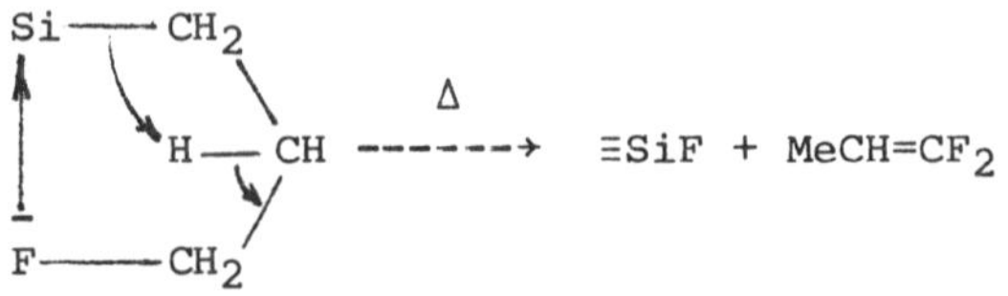

$$\equiv SiF + MeCH=CF_2 \qquad (86)$$

With siloxanes such as $(R_n^F R_{3-n}Si)_2O$ where $(R_n^F = CF_3CH_2CH_2)$, the C-Si bond fission is a homolytic process [181] that is understood to be due to a decrease of electrophilic character of silicon in these compounds. The 1,4-interaction between chlorine and silicon takes place [208] during the reaction of allyltrimethylsilane with the dimeric $[PtCl_4(C_2H_4)_2]$ (Eq. 87).

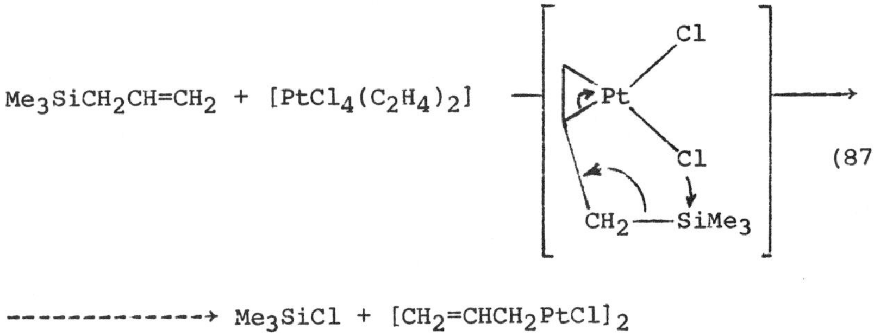

$$Me_3SiCH_2CH=CH_2 + [PtCl_4(C_2H_4)_2] \longrightarrow \left[\begin{array}{c} Cl \\ Pt \\ Cl \\ CH_2-SiMe_3 \end{array} \right] \longrightarrow \quad (87)$$

$$\dashrightarrow Me_3SiCl + [CH_2=CHCH_2PtCl]_2$$

2.3.4 Reactions Involving 1,5- or 1,6-Interaction

Interaction With Carbon. Both 1,5- and 1,6-interactions between silicon and a negatively charged carbon generally occur in a variety of ring-closure reactions [209] of difunctional organometallic reagents with silicon chlorides (Eq. 88),

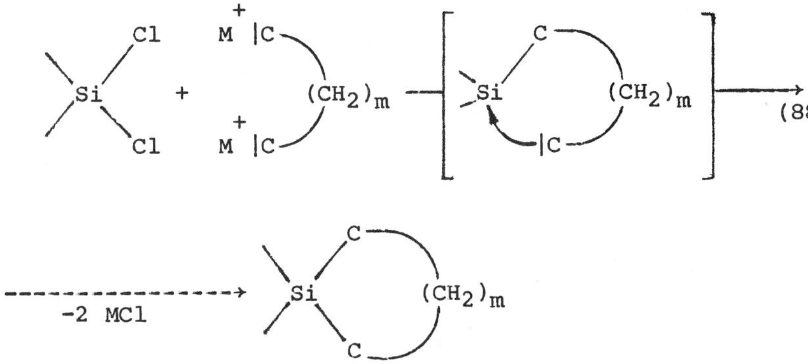

$$\begin{array}{c} Cl \\ Si \\ Cl \end{array} + \begin{array}{c} ^+M \; |C \\ (CH_2)_m \\ ^+M \; |C \end{array} \longrightarrow \left[\begin{array}{c} C \\ Si \quad (CH_2)_m \\ |C \end{array} \right] \longrightarrow \quad (88)$$

$$\underset{-2\;MCl}{\dashrightarrow} \quad \begin{array}{c} C \\ Si \quad (CH_2)_m \\ C \end{array}$$

or in cyclizations [190,209] of monofunctional organometallic silanes (Eq. 89).

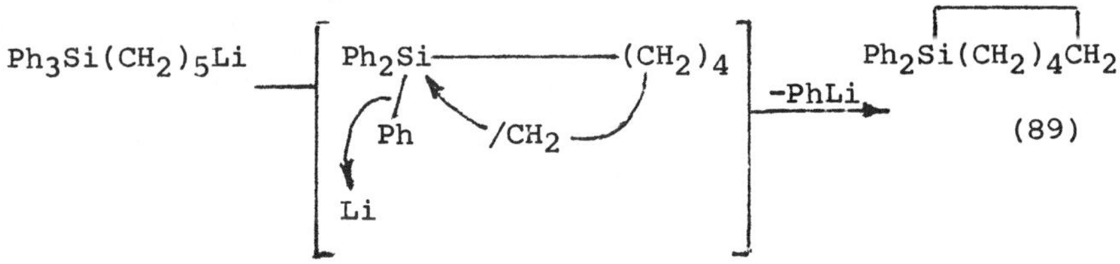

$$Ph_3Si(CH_2)_5Li \longrightarrow \left[\begin{array}{c} Ph_2Si-(CH_2)_4 \\ Ph \quad /CH_2 \\ Li \end{array} \right] \underset{-PhLi}{\longrightarrow} \begin{array}{c} \overline{Ph_2Si(CH_2)_4CH_2} \\ (89) \end{array}$$

The 1,5-anionic rearrangement of a trialkylsilyl group from carbon to carbon takes place in the reaction of N,N-dimethyl-2-triorganosilylethylamines with benzyne [188] (Eq. 90).

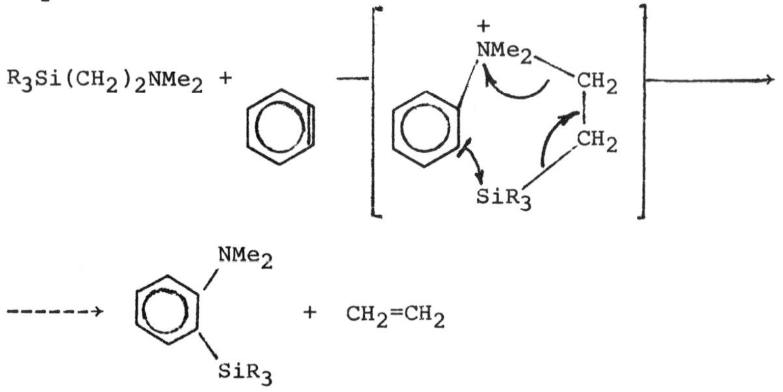

(90)

Interaction With Oxygen. The 1,5-rearrangement of a silyl group from carbon to oxygen proceeds during the interconversion of γ-silyl-α,β-unsaturated carbonyl compounds to siloxybutadiene [210] (Eq. 91).

(91)

In any migration of silicon from carbon to a carbonyl oxygen there are two effects working in opposition. The greater strength of Si-O bonds (~470 kJ/mol) relative to Si-C bonds (~310 kJ/mol) favors C→O migration, but the greater strength of the C=O bond relative to the C=C double bond favors O→C migration. For the unsaturated ketone (Eq. 91) the preference for silicon bonded to oxygen dominates, and its complete rearrangement to siloxybutadiene is observed. This 1,5-migration of silicon from carbon to oxygen is faster than the similar 1,3-migration.

The base-catalyzed cyclization of (4-ethoxysilylbutoxy)trimethylsilanes [200] has an identical course to the cyclization of (3-ethoxysilylpropoxy)trimethylsilanes (Section 2.3.3) and can therefore serve as an example of an intramolecular disproportionation initiated by the formation of a coordinative O→Si bond. Similarly, a 1,5- or 1,6- O→Si interaction can be presumed to occur in the following ring-closure reactions [211-214] (Eqs. 92-95).

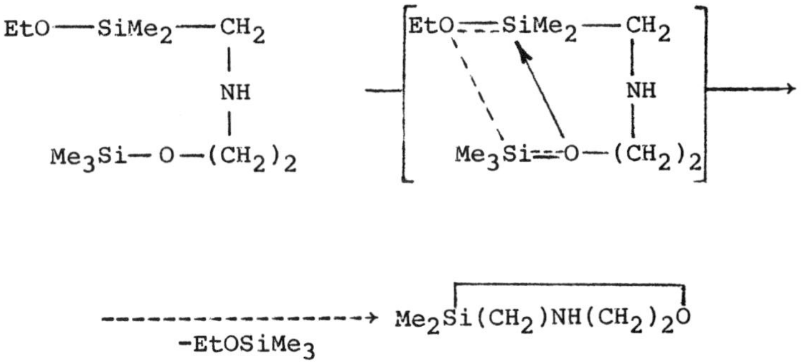

$$\begin{array}{c} \text{EtO—SiMe}_2\text{—CH}_2 \\ | \\ \text{NH} \\ | \\ \text{Me}_3\text{Si—O—(CH}_2)_2 \end{array} \longrightarrow \left[\begin{array}{c} \text{EtO}\cdots\text{SiMe}_2\text{—CH}_2 \\ | \\ \text{NH} \\ | \\ \text{Me}_3\text{Si}\cdots\text{O—(CH}_2)_2 \end{array} \right] \longrightarrow \quad (92)$$

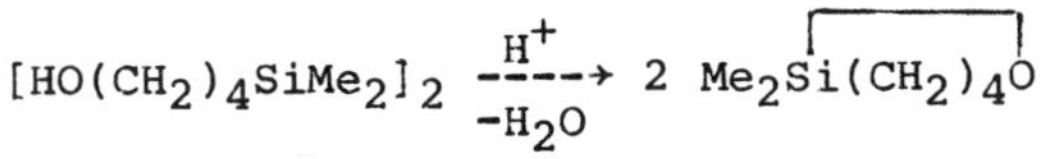

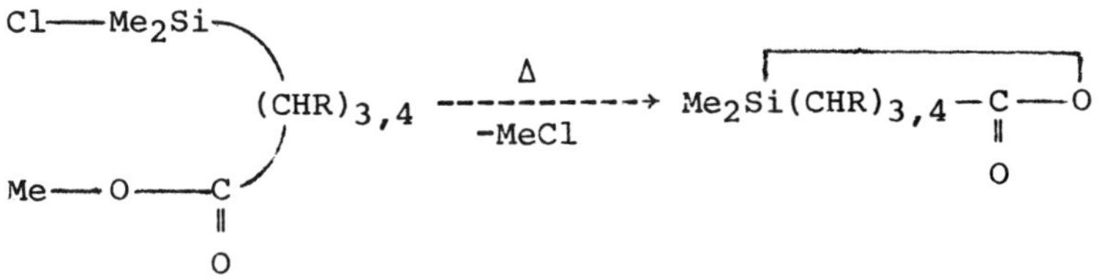

$$\xrightarrow[-\text{EtOSiMe}_3]{} \overline{\text{Me}_2\dot{\text{S}}\text{i(CH}_2)\text{NH(CH}_2)_2\dot{\text{O}}}$$

$$[\text{HO(CH}_2)_4\text{SiMe}_2]_2 \xrightarrow[-\text{H}_2\text{O}]{\text{H}^+} 2\ \overline{\text{Me}_2\dot{\text{S}}\text{i(CH}_2)_4\dot{\text{O}}} \quad (93)$$

$$\begin{array}{c} \text{Cl—Me}_2\text{Si} \\ \diagdown \\ (\text{CHR})_{3,4} \\ \diagup \\ \text{Me—O—C} \\ \| \\ \text{O} \end{array} \xrightarrow[-\text{MeCl}]{\Delta} \overline{\text{Me}_2\text{Si(CHR)}_{3,4}\text{—C—O}} \atop \| \atop \text{O} \quad (94)$$

$$\begin{array}{c} \text{EtO—SiMe}_2 \\ \diagdown \\ (\text{CH}_2)_4 \\ \diagup \\ \text{Me—C—O} \\ \| \\ \text{O} \end{array} \xrightarrow[-\text{MeCO}_2\text{Et}]{\text{base, }\Delta} \overline{\text{Me}_2\dot{\text{S}}\text{i(CH}_2)_4\dot{\text{O}}} \quad (95)$$

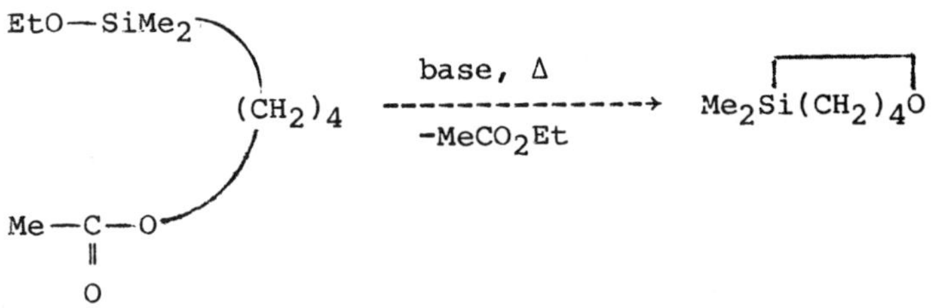

The following reactions involving a 1,5-interaction between a neutral silyl center and an oxygen of a functional group take place under electron impact in a mass spectrometer [215,216] (Eqs. 96,97)].

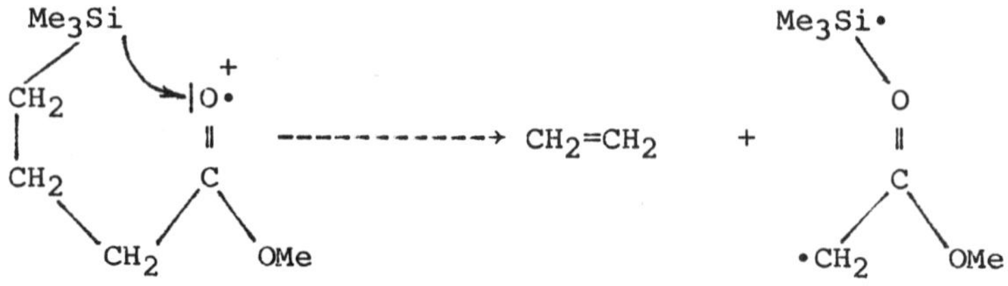

$$\text{(96)}$$

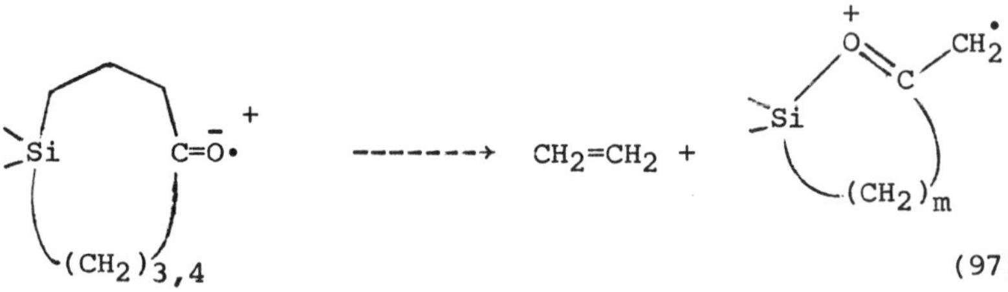

$$\text{(97)}$$

m=3,4

<u>Interaction With Halogens.</u> Interactions between silicon and a remote halogen are rare. They may be assumed to occur in the course of decomposition of diazonium salts [217] (Eq. 98),

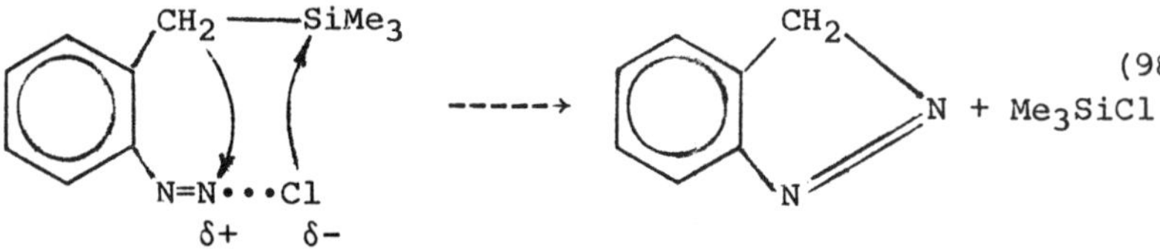

$$\text{(98)}$$

during decomposition [218] of 1-(chloromethyl)dimethylsilyl-3-trimethylsilylpropane in the presence of a catalytic amount of $AlCl_3$ (Eq. 99),

$$\text{(99)}$$

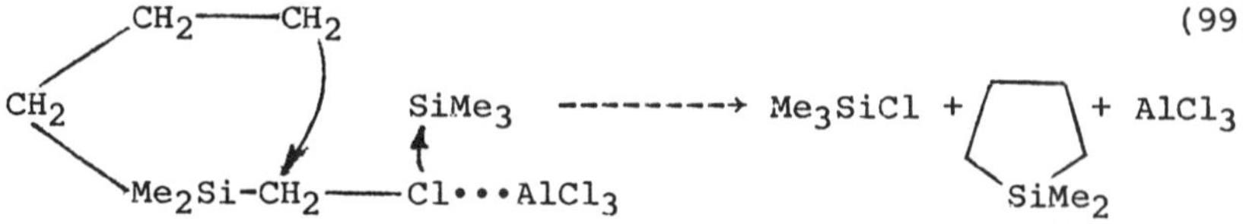

and during nucleophilic substitution of a hydroxyl group of 6,6-dimethyl-6-silacycloundecanols with a fluorine atom [219] (Eq. 100),

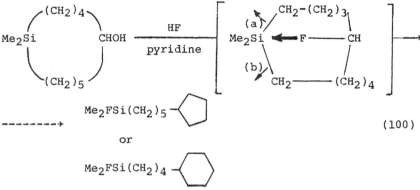

(100)

which gives a compound resulting from a transannular reaction at the silicon atom and cleavage of one intracyclic Si-C bond, which can take place either at (a) or at (b).

REFERENCES

1. Sommer L.H., Whitmore F.C.: J. Amer. Chem. Soc. 68, 485 (1946).
2. Sommer L.H., Baughman G.A.: J. Amer. Chem. Soc. 83, 3346 (1961).
3. Fleming I.: Chem. Ind. 1975, 449.
4. Eisch J.J., Trainor J.T.: J. Org. Chem. 28, 2870 (1963).
5. Litvinova O.V., Baukov Yu. I., Lutsenko I.F.: Dokl. Akad. Nauk USSR, 173, 578 (1967).
6. Ponomarev V.V., Golubcov S.A., Andrianov K.A., Kondrashova G.N.: Izv. Akad. Nauk USSR, Ser. Khim. 1969, 1545.
7. Davidson P.J., Lappert M.F., Pearce R.: Acc. Chem. Res. 7, 209 (1974).
8. Cundy C.S., Lappert M. F., Pearce R.: J. Organometal. Chem. 59, 161 (1973).
9. Eaborn C., Happer D.A.R., Safa K.D., Walton D.R.M.: J. Organometal. Chem. 157, C50 (1978).
10. Dua S.S., Eaborn C., Happer D.A.R., Hopper S.P., Safa K.D., Walton D.R.M.: J. Organometal. Chem. 178, 75 (1979).
11. Eaborn C., Retta N., Smith J.D.: J. Organometal. Chem. 190, 101 (1980).
12. Cook M.A., Eaborn C., Jukes A.E., Walton D.R.M.: J. Organometal. Chem. 24, 529 (1970).

13. Eaborn C., Eidenschink R., Jackson P.M., Walton D.R.M.: J. Organometal. Chem. 101, C 40 (1975).
14. Bourne A.J., Jarvie A.W.P., Holt A.: J. Chem. Soc. C, 1970, 1740.
15. Petrov A.D., Egorov Yu.P., Mironov V.F., Nikiskhin G.I., Bugorkova A.A.: Izv. Akad. Nauk USSR, Ser. Khim. 1956, 50.
16. Bugorkova A.A., Mironov V.F., Petrov A.D.: Izv. Akad. Nauk USSR, Ser. Khim. 1960, 474.
17. Ponomarev V.V., Golubcov S.A., Andrianov K.A., Chuprova E.A.: Izv. Akad. Nauk USSR, Ser. Khim. 1969, 1551.
18. Mironov V.F.: Izv. Akad. Nauk USSR, Ser. Khim. 1959, 1862.
19. Sommer L.H., Tyler L.J., Whitmore F.C.: J. Amer. Chem. Soc. 70, 2872 (1948).
20. Carsson E.: Trans. Chabmer. Univ. Technol. Gothenburg 115, 25 (1951).
21. Sakurai H., Hosomi A., Kumada M.: J. Org. Chem. 34, 1764 (1969).
22. Martin M.M., Gleicher G.J.: J. Amer. Chem. Soc. 86, 233, 238, 242 (1964).
23. Jarvie A.W.P., Rowley R.J.: J. Chem. Soc. B, 1971, 2439.
24. Cudlín J., Chvalovský V.: Coll. Czech. Chem. Commun. 27, 1658 (1962).
25. Dyakonov I.A., Golodnikov G.V., Repinskaya I.B.: Zh. Obshch. Khim. 35, 2181 (1965).
26. Dyakonov I.A., Repinskaya I.B., Golodnikov G.V.: Zh. Obshch. Khim. 36, 949 (1966).
27. Kirmse W.: Angew. Chem. 14, 678 (1962).
28. Chernyshev E.A.: Izv. Akad. Nauk USSR, Ser. Khim. 1960, 80.
29. Topchiev A.V., Nametkin N.S., Durgaryan S.G.: Zh. Obshch. Khim. 1960, 927.
30. Eaborn C., Emokpae T.A., Sidorov V.I., Taylor R.: J. Chem. Soc. Perkin II, 1974, 1454.
31. Bailey F.P., Taylor R.: J. Chem. Soc. B, 1971, 1446.
32. Bott R.W., Eaborn C., Leyshon K.: J. Chem. Soc. 1964, 1971.
33. Walton D.R.M.: J. Organometal. Chem. 3, 438 (1965).
34. Bassindale A.R., Eaborn C., Walton D.R.M., Young D.J.: J. Organometal. Chem. 20, 49 (1969).
35. Chernyshev E.A., Dolgaya M.E., Petrov A.D.: Izv. Akad. Nauk USSR, Ser. Khim. 1960, 1424.
36. Chernyshev E.A., Tolstikova N.G.: Izv. Akad. Nauk USSR, Ser. Khim. 1960, 1595.

37. Chernyshev E.A., Klyukina E.N., Petrov A.D.: Izv. Akad. Nauk USSR, Ser. Khim. 1960, 1601.
38. Eaborn C.: J. Chem. Soc. 1956, 4858.
39. Traylor T.G., Hanstein W., Berwin H.J., Clinton N.A., Brown R.S.: J. Amer. Chem. Soc. 93, 5715 (1971).
40. Hanstein W., Berwin H.J., Traylor T.G.: J. Amer. Chem. Soc. 92, 829 (1970).
41. Eaborn C., Parker S.H.: J. Chem. Soc. 1954, 939.
42. Cook M.A., Eaborn C., Walton D.R.M.: J. Organometal. Chem. 24, 293 (1970).
43. Benkeser R.A., Bennett E.W.: J. Amer. Chem. Soc. 80, 5414 (1958).
44. Sommer L.H., Gold J.R., Goldberg G.M., Marans N.S.: J. Amer. Chem. Soc. 71, 1509 (1949).
45. Rijkens F., Jansen M.J., Drenth W., Van Denkerk G.J.M.: J. Organometal. Chem. 2, 347 (1964).
46. Sommer L.H., Rockett J.: J. Amer. Chem. Soc. 73, 5130 (1951).
47. Voronkov M.G., Kashik T.V., Lukevics E.Y., Deriglazova E.S., Pestunovich A.E., Moskovich R.Y.: Zh. Obshch. Khim. 44, 778 (1974).
48. Voronkov M.G., Kashik T.V., Deriglazova E.S., Lukevics E.Y., Pestunovich A.E., Sturkovich R.Y.: Zh. Obshch. Khim. 46, 1522 (1976).
49. Voronkov M.G., Kashik T.V., Lukevics E.Y., Deriglazova E.S., Pestunovich A.E.: Zh. Obshch. Khim. 45, 2200 (1975).
50. Andrianov K.A., Kopylov V.M., Chernyshev A.I., Andreva S.V., Shragin I.S.: Zh. Obshch. Khim. 45, 351 (1975).
51. Fialová V., Bažant V., Chvalovský V.: Coll. Czech. Chem. Commun. 38, 3837 (1973).
52. Lukevics E.Y., Voronkov M.G., Shestakov E.E., Pestunovich A.E.: Zh. Obshch. Khim. 41, 2218 (1971).
53. Shea K.J., Gobeille R., Bramblett J., Thompson E.: J. Amer. Chem. Soc. 100, 1611 (1978).
54. Egorochkin A.N., Skobeleva S.E., Sevastyanova E.I., Kosolapova I.G., Sheludyakov V.D., Rodionov E.S., Kirilin A.D.: Zh. Obshch. Khim. 46, 1795 (1976).
55. Pola J., Bažant V., Chvalovský V.: Coll. Czech. Chem. Commun. 37, 3885 (1972).
56. Jarvie A.W.P., Holt A., Thompson J.: J. Chem. Soc. B 1969, 852.
57. Brownbridge P., Fleming I., Pearce A., Warren S.: J. Chem. Soc., Chem. Commun. 1976, 751.
58. Davidson A.H., Fleming I., Grayson J.I., Pearce A., Snowden R.L., Warren S.: J. Chem. Soc. Perkin I, 1977, 550.

59. Fleming I., Patel S.K.: Tetrahedron Lett. 22, 2321 (1981).
60. Eaborn C., Mahmoud F.M.S.: J. Organometal. Chem. 209, 13 (1981).
61. Davis D.D., Black R.H.: J. Organometal. Chem. 82, C 30 (1974).
62. Washburne S.S., Chawla R.R.: J. Organometal. Chem. 133, 7 (1977).
63. Heck R., Prelog V.: Helv. Chim. Acta 38, 1541 (1955).
64. Pola J., Chvalovský V.: Coll. Czech. Chem. Commun. 39, 2247 (1974).
65. Pola J., Chvalovský V.: Coll. Czech. Chem. Commun. 39, 2637 (1974).
66. Shostakovskii M.F., Shikhiev I.A., Komarov N.V.: Izv. Akad. Nauk USSR, Ser. Khim. 1956, 1493.
67. Shostakovskii M.F., Vlasova N.N., Cyganskaya I.I., Emelyanov I.F.: Zh. Obshch. Khim. 40, 1897 (1978).
68. Pola J., Bellama J.M., Chvalovský V.: Coll. Czech. Chem. Commun. 39, 3705 (1974).
69. Badger R.M., Bauer J.M.: J. Chem. Phys. 5, 839 (1937).
70. Drago R.S.: Structure and Bonding 15, 73 (1973).
71. Pola J., Jakoubková M., Chvalovský V.: Coll. Czech. Chem. Commun. 43, 760 (1978).
72. Pola J., Jakoubková M., Chvalovský V.: Coll. Czech. Chem. Commun. 40, 2063 (1975) and references therein.
73. Pola J., Chvalovský V.: Coll. Czech. Chem. Commun. 38, 1674 (1973).
74. Bellama J.M., Davison J.B.: Inorg. Chem. 14, 3119 (1975).
75. Bellama J.M., Harmon L.A.: Inorg. Chem. 17, 482 (1978).
76. Bellama J.M., Gerchman L.L.: Inorg. Chem. 14, 1618 (1975).
77. Huheey J.E.: J. Phys. Chem. 69, 3284 (1965).
78. Wang J.T., Van Dyke C.H.: Inorg. Chem. 6, 1741 (1967).
79. Gibbon G.A., Wang J.T., Van Dyke C.H.: Unpublished results, cited in ref. 76.
80. Pola J., Chvalovský V.: Coll. Czech. Chem. Commun. 45, 861 (1980).
81. Hetflejš J., Svoboda L., Jakoubková M., Chvalovský V.: Coll. Czech. Chem. Commun. 38, 717 (1973).
82. Shostakovskii M.F., Komarov N.V., Vlasova N.N.: Zh. Obshch. Khim. 37, 1151 (1967).
83. Shostakovskii M.F., Komarov N.V., Vlasova N.N., Rinkus G.A.: Zh. Obshch. Khim. 36, 904 (1966).

84. Ingold C.K.: Structure and Mechanism in Organic Chemistry, Second Ed., Cornell University Press, Ithaca 1969, Chapter 7.
85. Eaborn C.: Organosilicon Compounds, Butterworths, London 1960, p. 433.
86. Cook M.A., Eaborn C., Walton D.R.M.: J. Organometal. Chem. 29, 389 (1971).
87. Cartledge F.K., Jones J.P.: Tetrahedron Lett. 1971, 2193.
88. Cartledge F.K., Jones J.P.: J. Organometal. Chem. 67, 379 (1974).
89. Cook M.A., Eaborn C., Walton D.R.M.: J. Organometal. Chem. 24, 301 (1970).
90. Eaborn C., Jeffrey J.C.: J. Chem. Soc. 1954, 4266.
91. Cooper G.D., Prober M.: J. Amer. Chem. Soc. 76, 3943 (1954).
92. Mironov V.F., Pogonkina N.A.: Izv. Akad. Nauk USSR, Ser. Khim. 1957, 1199.
93. Huang C.T., Wang P.J.: Hua Hsüeh Hsüeh Pao 25, 330 (1959), Chem. Abstr. 54, 16375 d (1960).
94. Eaborn C., Jeffrey C.: J. Chem. Soc. 1957, 137.
95. Whitmore F.C., Sommer L.H.: J. Amer. Chem. Soc. 68, 481 (1946).
96. Miller V.B., Neiman M.B., Savickii A.V., Mironov V.F.: Dokl. Akad. Nauk USSR, 121, 495 (1955).
97. Smith R.P., Eyring H.: J. Amer. Chem. Soc. 74, 229 (1952).
98. Boak D.S., Gowenlock B.G.: J. Organometal. Chem. 29, 385 (1971).
99. Včelák J., Roman L., Chvalovský V.: Coll. Czech. Chem. Commun. 41, 2708 (1976).
100. Ref. 83, pp 204-206.
101. Steward O.W., Pierce O.R.: J. Amer. Chem. Soc. 83, 4932 (1961).
102. Joesten M.D., Shaad L.J.: Hydrogen Bonding, Chap. 4, M. Dekker. New York 1974.
103. Badger R.M., Bauer J.M.: J. Chem. Phys. 5, 835 (1937).
104. Drago R.S.: Structure and Bonding 15, 73 (1973).
105. Pola J., Jakoubková M., Chvalovský V.: Coll. Czech. Chem. Commun. 43, 753 (1978).
106. Pola J., Chvalovský V.: Coll. Czech. Chem. Commun. 43, 3192 (1978).
107. Zhdanov Y.A., Minkin V.I.: Korrelacionnyi Analiz v Organischeskoi Khimii, Izd. Rostov. Univ., Rostov 1966.
108. Voronkov M.G., Feshin V.P., Mironov V.F., Mikhailants S.A., Gar T.K.: Zh. Obshch. Khim. 41, 2211 (1971).

109. Pola J., Chvalovský V.: Coll. Czech. Chem. Commun. 42, 3581 (1977).
110. Chernyshev E.A., Dolgaya M.E.: Zh. Obshch. Khim. 29, 1850 (1959).
111. Abel E.W., Dunster M.O., Waters A.: J. Organometal. Chem. 49, 287 (1973).
112. Larrabee R.B.: J. Organometal. Chem. 74, 313 (1974).
113. Asche A.J.: J. Amer. Chem. Soc. 92, 1233 (1970).
114. Fritz H.P., Kreiter C.G.: J. Organometal. Chem. 4, 313 (1965).
115. Calderon J.L., Cotton F.A., Legzdins P.: J. Amer. Chem. Soc. 91, 2528 (1969).
116. Sergeyev N.M., Avramenko G.I., Ustynuk Y.A.: J. Organometal. Chem. 22, 79 (1970).
117. Egger K.W., James T.L.: J. Organometal. Chem. 26, 335 (1971).
118. Sergeyev N.M., Avramenko G.I., Kisin A.V., Korenvesky U.A., Ustynuk Y.A.: J. Organometal. Chem. 32, 55, 77 (1971).
119. Luzikov Y.N., Sergeyev N.M., Ustynuk Y.A.: J. Organometal. Chem 65, 303 (1974).
120. Boche G., Heidenhaim F.: J. Organometal. Chem. 121, C 49 (1976).
121. Mironov V.F., Sherbinin V.V., Viktorov N.A., Sheludyakov V.D.: Zh. Obshch. Khim. 45, 1908 (1975).
122. Asche A.J.: J. Amer. Chem. Soc. 95, 818 (1973).
123. Cunico R.F., Lee H.M.: J. Amer. Chem. Soc. 99, 7613 (1977).
124. Reetz M.T.: Tetrahedron 29, 2189 (1973).
125. Nametkin N.S., Grushevenko I.A., Perchenko V.N., Kamneva G.L.: Dokl. Akad. Nauk USSR 207, 865 (1972).
126. Brook A.G., Duff J.M.: J. Amer. Chem. Soc. 96, 4692 (1974).
127. Bassindale A.R., Brook A.G., Jones D.F., Stewart J.A.G.: J. Organometal. Chem. 152, C 25 (1978).
128. Kaufmann K.D., Rühlmann K.: Z. Chem. 8, 262 (1968).
129. Brook A.G.: Acc. Chem. Res. 7, 77 (1974).
130. Gilman H., Wu T.C.: J. Amer. Chem. Soc. 75, 2935 (1953).
131. Brook A.G., Limburg W.W., MacRae D.M., Fieldhouse S.A.: J. Amer. Chem. Soc. 89, 704 (1967).
132. Brook A.G., Schwartz N.V.: J. Org. Chem. 27, 2311 (1962).
133. Brook A.G., Fieldhouse S.A.: J. Organometal. Chem. 10, 235 (1967).

134. Bassindale A.R., Brook A.G., Chen P., Lennon J.: J. Organometal. Chem. 94, C 21 (1975).
135. Wilt J.W., Kolewe O., Kraemer J.F.: J. Amer. Chem. Soc. 91, 2624 (1969).
136. Reetz M.T.: Angew. Chem., Int. Ed. Engl. 13, 402 (1974).
137. Atwell W.H., Weyenberg D.R., Uhlmann J.G.: J. Amer. Chem. Soc. 91, 2024 (1969).
138. Haszeldine R.N., Rogers D.J., Tipping A.E.: J. Organometal. Chem. 54, C 5 (1973).
139. Brook A.G., Jones P.F.: J. Chem. Soc., Chem. Commun. 1969, 1325.
140. Wright A., West R.: J. Amer. Chem. Soc. 96, 3222 (1974).
141. Fishwick G., Haszeldine R.N., Parkinson C., Robinson P.J., Simmons R.F.: J. Chem. Soc., Chem. Commun. 1965, 382.
142. Haszeldine R.N., Young J.C.: Proc. Chem. Soc. 1959, 394.
143. Weber W.P., Felix R.A., Willard A.K., Koenig K.E.: Tetrahedron Lett. 1971, 4701.
144. Bevan W.I., Haszeldine R.N., Middleton J., Tipping A.E.: J. Organometal. Chem. 23, C 17 (1970).
145. Whitmore F.C., Sommer L.H., Gold J.: J. Amer. Chem. Soc. 69, 1976 (1947).
146. O'Brien D.H., Hairston T.J.: Organometal. Chem. Rev. A 7, 132 (1971).
147. Bott R.W., Eaborn C., Rushton B.M.: J. Organometal. Chem. 3, 455 (1965).
148. Bellama J.M., MacDiarmid A.G.: J. Organometal. Chem. 18, 275 (1969).
149. Voronkov M.G., Kirpichenko S.V., Keiko V.V., Pestunovich V.A., Tsetlina E.O., Chvalovský V., Včelák J.: Izv. Akad. Nauk USSR, Ser. Khim. 1975, 2052.
150. Corey J.Y., Chang V.H.T.: J. Organometal. Chem. 174, C 15 (1979).
151. Slutsky J., Kwart H.: J. Amer. Chem. Soc. 95, 8678 (1973).
152. Ishikawa M., Sugisawa H., Fábry L., Kumada M.: J. Organometal. Chem. 161, 299 (1978).
153. Belavin I.Y., Fedoseeva N.A., Baukov Y.I., Lutsenko I.F.: Zh. Obshch. Khim. 43, 433 (1973).
154. Brook A.G., MacRae D.M., Limburg W.W.: J. Amer. Chem. Soc. 89, 5493 (1967).
155. Kwart H., Barnette W.E.: J. Amer. Chem. Soc. 99, 614 (1977).

156. Larson G.L., Fernandez Y.V.: J. Organometal. Chem. 86, 193 (1975).
157. Brook A.G., Anderson D.F., Duff J.M.: J. Amer. Chem. Soc. 90, 3876 (1968).
158. Brook A.G., Duff J.M., Anderson D.G.: J. Amer. Chem. Soc. 92, 7567 (1970).
159. Brook A.G., Anderson D.G.: Can. J. Chem. 46, 2115 (1968).
160. Vedejs E., Mullins M.: Tetrahedron Lett. 1975, 2017.
161. Nesmeyanov A.N.: J. Organometal. Chem. 100, 161 (1975).
162. Lutsenko I.F., Baukov Y.I., Kostyuk A.S., Savelyeva N.I., Krysina V.K.: J. Organometal. Chem. 17, 241 (1969).
163. Itoh K., Kato S., Ishii Y.: J. Organometal. Chem. 34, 293 (1973).
164. Canavan A.E., Eaborn C.: J. Chem. Soc. 1962, 592.
165. Saveleva N.I., Kostyuk A.S., Baukov Y.I., Lutsenko I.F.: Zh. Obshch. Khim. 41, 485 (1971).
166. Musker W.K., Larson G.L.: J. Amer. Chem. Soc. 91, 514 (1969).
167. Gilman H., Tomasi R.A.: J. Org. Chem. 27, 3647 (1962).
168. Connor J.A., Jones E.M.: J. Organometal. Chem. 60, 77 (1973).
169. Peterson D.J.: J. Org. Chem. 33, 780 (1968).
170. Chan T.H., Chang E., Vinokur E.: Tetrahedron Lett. 1970, 1137.
171. Matsuda I., Izumi Y.: Tetrahedron Lett. 22, 1805 (1981).
172. Seyferth D., Lefferts J.L., Lambert R.L.: J. Organometal. Chem. 142, 39 (1977).
173. Seebach D., Gröbel B.T., Beck A.K., Braun M., Geiss K.H.: Angew. Chem., Int. Ed. Engl. 11, 443 (1972).
174. Carey F.A., Hernandez O.: J. Org. Chem. 38, 2670 (1973).
175. Shimoji K., Taguchi H., Oshima K., Aymamoto H., Nozaki H.: J. Amer. Chem. Soc. 96, 1620 (1974).
176. Haszeldine R.N., Robinson P.J., Simmons R.F.: J. Chem. Soc. 1964, 1890.
177. Graham D., Haszeldine R.N., Robinson P.J.: J. Chem. Soc. B 1969, 652.
178. Graham D., Haszeldine R.N., Robinson P.J.: J. Chem. Soc. B 1971, 611.
179. Bell T.N., Berkley R., Platt A.E., Sherwood A.G.: Canad. J. Chem. 52, 3158 (1974).
180. Fishwick G., Haszeldine R.N., Robinson P.J., Simmons R.F.: J. Chem. Soc. B 1970, 1236.

181. Bell T.N., Haszeldine R.N., Newlands M.J., Plumb J.B.: J. Chem. Soc. 1965, 2107.
182. Davidson I.M.T., Eaborn C., Lilly M.N.: J. Chem. Soc. 1964, 2624, 2630.
183. Davidson I.M.T.: J. Chem. Soc. 1965, 5481.
184. Davidson I.M.T., Jones M.R., Pett C.: J. Chem. Soc. B 1967, 937.
185. Ponomarev V.V., Shapatin A.S., Golubcov S.A.: Zh. Obshch. Khim. 36, 364 (1966).
186. Sato Y., Ban Y., Shirai H.: J. Organometal. Chem. 113, 115 (1976).
187. Sato Y., Toyooka T., Aoyama T., Shirai H.: J. Org. Chem. 41, 3559 (1976).
188. Sato Y., Ban Y, Aoyama T., Shirai H.: J. Organometal. Chem. 41, 1962 (1976).
189. Gilman H., Tomasi R.A.: J. Amer. Chem. Soc. 81, 137 (1959).
190. Maercker A., Eckers M., Passlack M.: J. Organometal. Chem. 186, 193 (1980).
191. Dolzine T.W., Hovland A.K., Oliver J.P.: J. Organometal. Chem. 65, C 1 (1974).
192. Scherer O.J., Schnabl G.: J. Organometal. Chem. 52, C 18 (1973).
193. Belyakova Z.V., Bochkarev V.N., Golubcov S.A., Belikova Z.V., Yamova M.S., Ainshtein A.A., Baranova G.G., Efremova L.A., Popkov K.K.: Zh. Obshch. Khim. 42, 858 (1972).
194. Bailey D.L.: U.S. Patent 2 888 454 (1959).
195. Andrianov K.A., Pakhomov V.I., Lapteva N.E.: Dokl. Akad. Nauk USSR, 151, 849 (1963).
196. Manuel G., Mazerolles P., Florence J.C.: J. Organometal. Chem. 30, 5 (1971).
197. Massol M., Barrau J., Satge J., Bouyssieres B.: J. Organometal. Chem. 80, 47 (1974).
198. Speier J.L., David M.P., Eynon B.A.: J. Org. Chem. 25, 1637 (1960).
199. Sommer L.H.: U.S. Patent 2 589 446 (1952).
200. Pola J., Bažant V., Chvalovský V.: Coll. Czech. Chem. Commun. 38, 1528 (1973).
201. Grignon-Dubois M., Dunogues J., Duffant N., Calas R.: J. Organometal. Chem. 188, 311 (1980).
202. Tsuge O., Sone K., Urano S., Matsuda K.: J. Org. Chem. 47, 5171 (1982).
203. Bac N.V., Langlois Y.: J. Amer. Chem. Soc. 104, 7666 (1982).

204. Onan K.D., McPhail A.T., Yoder C.H., Hilliard R.W.: J. Chem. Soc., Chem. Commun. 1978, 209.
205. Hilliard R.W., Ryan C.M., Yoder C.H.: J. Organometal. Chem. 153, 369 (1978).
206. Weber W.P., Felix R.A., Willard A.K., Boettger H.G.: J. Org. Chem. 36, 4060 (1971).
207. Novikov S.N., Kazan E.G., Pravednikov A.N.: Zh. Obshch. Khim. 38, 402 (1968).
208. Mansuy D., Pusset J., Chottard J.C.: J. Organometal. Chem. 110, 139 (1976).
209. Wittenberg D., Gilman H.: J. Amer. Chem. Soc. 80, 2677 (1958) and references therein.
210. Casey C.P., Jones C.R., Tukada H.: J. Org. Chem. 46, 2089 (1981).
211. Zhdanov A.A., Pakhomov V.I., Arkhipov I.A.: Plast. Massy 1966, 19.
212. Simler W., Niederprüm H., Sattlegger H.: Chem. Ber. 99, 1368 (1966).
213. Saam J.C.: U.S. Patent 3 395 167 (1968).
214. Pola J., Jakoubková M., Chvalovský V.: Coll. Czech. Chem. Commun. 41, 374 (1976).
215. Weber W.P., Felix R.A., Willard A.K.: J. Amer. Chem. Soc. 92, 1420 (1970).
216. Weber W.P., Felix R.A., Wilard A.K., Boettger H.G.: J. Org. Chem. 36, 4060 (1971).
217. Benkeser R., Brumfield P.E.: J. Amer. Chem. Soc. 74, 253 (1952).
218. Nametkin N.S., Vdovin V.M., Pushevaya K.S.: Dokl. Akad. Nauk USSR 150, 562 (1963).
219. Duboudin F., Bourgeois G., Faucher A., Mazerolles P.: J. Organometal. Chem. 133, 29 (1977).

3

NMR SPECTROSCOPY IN THE INVESTIGATION AND ANALYSIS OF

CARBON-FUNCTIONAL ORGANOSILICON COMPOUNDS

Jan Schraml

Institute of Chemical Process Fundamentals
Czechoslovak Academy of Sciences
Prague, Czechoslovakia

3.1 INTRODUCTION

A decade ago, a review of the literature on organosilicon compounds [1], which covered approximately 15,000 papers since 1823, identified only some 1,400 papers on the NMR spectroscopy of organosilicon compounds. Of these, only a few were concerned with carbon-functional organosilicon compounds. It was obvious that rapid progress was to be expected in this field, and the past decade has indeed shown a considerable increase in interest in carbon-functional compounds, an interest that is also documented in the other chapters of this volume.

This increased interest came hand-in-hand with rapid progress in NMR experimental techniques. Using pulsed Fourier transform NMR spectrometers, the spectra of all NMR-active nuclei in carbon-functional compounds can now be measured, and thus it is now possible to detect various inter- or intra-molecular interactions. The presence of intervening carbon atoms will allow the separation of different mechanisms that substituent effects have on chemical shifts or coupling constants, and thus carbon-functional compounds become convenient model compounds on which spectroscopists test proposed theories of shielding or coupling. On the other hand, high-field spectrometers with better spectral dispersion and higher sensitivity produce spectra that are easier to interpret and thus make NMR spectroscopy the analytical method of choice for the synthetic chemist. Naturally, the combined outcome of these trends is a broader range of problems that can now be studied in greater depth.

The scope of the present review concentrates on high-resolution studies of liquid samples; relaxation studies are not considered. An exhaustive review of the present theme is not feasible; hence, a representative approach is adopted. The classes of carbon-functional compounds covered is also of necessity limited.

Carbon-functional compounds can be viewed as derived from (parent) organosilicon compounds (type A) by replacing a carbon-bound hydrogen atom with functional group Y (type B):

$$X^1X^2X^3Si-(CH_p)_m-H \qquad X^1X^2X^3Si-(CH_p)_m-Y$$

$$\text{type A} \qquad\qquad\qquad \text{type B}$$

Of concern here are only the carbon-functional compounds of type B (i.e., those that have only one functional group Y) and that have the connecting chain $(CH_p)_m$ either aliphatic, $(CH_2)_m$, or aromatic, (C_6H_4). These two classes of carbon-functional compounds will be called aliphatic and aromatic, respectively.

Aliphatic and aromatic carbon-functional compounds represent two distinctly different situations: in aliphatic compounds the behavior of various silyl groups is usually commensurate with the electronegativity and polarizability of the silicon atom; in aromatic compounds silicon behaves as a weak electron acceptor. Studies of aliphatic carbon-functional compounds are complicated by their high conformational mobility.

In general, analysis and investigation of carbon-functional organosilicon compounds by NMR spectroscopy can utilize nuclei of ^1H, ^{13}C, and ^{29}Si, and also other NMR-active nuclei present in the silicon substituents (X) or in the carbon-functional groups (Y).

Within the scope of this chapter even a brief review of the NMR spectroscopy of all of these nuclei is not possible. It is assumed that the reader is familiar with ^1H and ^{13}C NMR spectroscopy. Extensive reviews of the NMR of "other" nuclei have been published recently [2]. The field of ^{29}Si-NMR has also been reviewed by several authors [3-6] in considerable detail. However, since silicon plays a key role in the compounds considered here, and since to many readers this branch of NMR spectroscopy is not yet overly familiar, the main features of ^{29}Si-NMR spectroscopy will be summarized here.

3.2 ^{29}Si NMR SPECTROSCOPY

3.2.1 Experimental Aspects

Sensitivity. ^{29}Si is the only magnetically active natural isotope of silicon. It has spin I = 1/2, a magnetogyric ratio $\gamma = -5.314 \times 10^{-7}$ rad $T^{-1}s^{-1}$, and a natural abundance of 4.70%, which give a theoretical NMR receptivity of 3.69×10^{-4} that of ^{1}H, and 2.1 that of ^{13}C [5]. In practice, the efficiency of ^{29}Si NMR measurements is reduced by two factors: (a) the negative magnetogyric ratio, which often leads to a negative nuclear Overhauser effect [7] in proton decoupled spectra*, and (b) ^{29}Si nuclei tend to have long spin-lattice relaxation times.

Indirect Observation of ^{29}Si Resonance. Indirect observation is achieved through heteronuclear X-$\{^{29}$Si$\}$ double resonance experiments by which spectra of nuclei X with higher NMR sensitivity are measured (X = ^{1}H or ^{19}F). Methods that can serve this purpose have been reviewed by

---

*The nuclear Overhauser effect (NOE) is the line intensity change upon saturation of proton resonance. The observed NOE is proportional to the ratio of the total T_1 relaxation time of silicon to the relaxation time from the dipole-dipole relaxation mechanism. When silicon relaxation is dominated by the dipole-dipole mechanism, the ^{29}Si line observed with proton decoupling is inverted, but it is enhanced by NOE by a factor of 1.5. At the other extreme, if the dipole-dipole mechanism does not contribute to relaxation, the NOE is zero and the intensity is unaltered (except for the collapse of the multiplet due to proton decoupling). Intermediate cases can lead to intensity loss and eventually to a null signal.

McFarlane [8]. Examples of their application to organosilicon compounds can be found in the early work of Johannesen et al. [9,10] and others [11-13], and more recently in the studies of the Pestunovich group [14,15]. The main advantages of these methods are the high sensitivity, precision, and relative ease (and low cost) of their implementation on CW spectrometers. They can be applied only to compounds with observable spin-spin coupling of the observed nucleus to ^{29}Si.

Direct Observation of ^{29}Si Resonance. Continuous-wave (CW) methods (for details of their application to organosilicon compounds, see [16,17]) appear to be obsolete now, but since practically all chemical shifts determined directly prior to 1974 were measured by these techniques, a few words of comment are in order. The low sensitivity required large samples, and compounds were usually measured as neat liquids using "external" references. Hence, the chemical shifts determined in this way are reproducible to within ±0.3 ppm at best, but in some instances systematic errors as large as 2 ppm were noted.

The pulsed Fourier transform method (FT NMR) is now preferred for reasons explained in every standard text [18,19], and peculiarities of ^{29}Si FT NMR are discussed in several specialized reviews [3-6,20,21]. Two techniques are now routine for overcoming the two factors that reduce the efficiency of "normal" measurements. These are: (a) gated proton decoupling, and (b) the use of (shiftless) relaxation reagents.

In the gated decoupling technique the protons are irradiated only through the period of acquisition of ^{29}Si FID data points. Short acquisition times (a few seconds, as dictated by the required resolution, memory size, and spectral width [18,19]) and long waiting periods between the excitation pulses do not allow a buildup of the nuclear Overhauser effect that can reduce the intensity gain due to collapse of the ^{29}Si multiplets by proton decoupling. Thus, this technique removes the effect of the negative NOE, but the measuring time is still lengthy because of the problem of slow relaxation.

Addition of a relaxation reagent to the solution shortens the relaxation time, thus making faster pulse repetition possible. Also, as the total relaxation time is shortened while relaxation by the dipole-dipole mechanism is not affected, the NOE is practically eliminated. The presence of a relaxation reagent also makes the T_1 relaxation times of all ^{29}Si nuclei in the sample equal, and the line intensities can then be utilized for quantitative studies. The most frequently used relaxation reagents are iron (III) or chromium (III) acetyl acetonates [20,21] (in 10^{-2} to 10^{-3} M concentrations). Relaxation reagents are not always without effect on chemical shifts [22], and in some cases can act as catalysts (e.g., perhalosilanes were decomposed by Cr(acac)$_3$ [23]). Of course, the sample is also contaminated when relaxation reagents are used.

Recent advances in NMR instrumentation have made possible the introduction of several new sophisticated methods for sensitivity enhancement. They are all based on a transfer of the large magnetization of protons to nuclei with low magnetogyric ratios. Since all of these methods rely on the scalar coupling of protons with the nuclei in question (which need not necessarily be one-bond coupling), they are not likely to find widespread use in routine applications. For series of similar compounds with similar couplings to protons, the use of such methods represents considerable savings in spectrometer time. Listed below are the most promising techniques (which are known by their acryonyms) with the appropriate references: AJCP (adiabatic J cross-polarization) [24], CP (cross-polarization) [25-31], DSPI (double selective population inversion) [32], INEPT (insensitive nuclei enhanced by polarization transfer) [33-36], PREP (population redistribution for enhancement with proton decoupling) [37], PSSPI (progressive saturation SPI) [38,39], SPI (selective population inversion) [40], SPT (selective population transfer) [23,25,41].

Heteronuclear spin-spin coupling constants of silicon-29 are usually much easier to measure from the spectra of the other involved nucleus with a higher NMR sensitivity. Homonuclear ^{29}Si-^{29}Si couplings must, however, be measured from ^{29}Si satellites in the ^{29}Si-NMR spectra. In natural abundance, ^{29}Si satellites (originating in the isotopomers with two ^{29}Si atoms) to the main ^{29}Si line (from the isotopomer with only one ^{29}Si atom) constitute only 4.70 $\underline{n}$ % of the main line ($\underline{n}$ is the number of chemically equivalent Si atoms). Though a suf-

Table I. Chemical shifts of ^{29}Si reference materials

Compound	Symbol	^{29}Si Chemical Shift, (external ref.)[b]	δ-Scale (internal ref.)	Solvent[a]
$(CH_3)_4Si$	TMS	0.0[c]	0.00[c]	any
$(C_2H_5O)_4Si$	Si(OEt)$_4$	-83.2[d]	-82.59[e]	CCl_4
$(CH_3O)_4Si$	TMOS	-79.5[f] -78.50[h]	-79.22[g] -78.48[i]	CCl_4 $CDCl_3$
$[(CH_3)_2SiO]_4$	D_4	-20.0[j]	-19.54[g] -19.51[k]	CCl_4 $(CD_3)_2CO$
$[(CH_3)_2SiO]_x$	D_x	-22.0[l]	-22.22[g,m]	CCl_4
$[(CH_3)_2SiO]_2O$	MM, HMDSO	6.3[j]	6.87[g] 6.79[k]	CCl_4 $(CD_3)_2CO$
$[(CH_3)_3SiO]_4Si$[n]		-105.2[j]		
$[(CH_3)_3Si]_2$	HMDSS	-19.58[g]	-19.77[o] -19.75[o] -19.82[o] -19.76[o] -19.79[o]	C_6H_{12} $CHCl_3$ dioxane CCl_4 $(CD_3)_2CO$

a The solvent refers to δ (internal reference) values.
b Usually measured in neat liquids.
c Primary reference.
d Data from Magi et al. [47], ±0.3 ppm; values between −81.0 and −83.5 ppm [17,48,49].
e Data from Levy et al. [50], ±0.07 ppm.
f Data from Hunter and Reeves [48], ±1 ppm; this value should be used for conversion of data.
g Value converted from data of Levy et al. [50], ±0.14 ppm.
h Data from Scholl et al. [51], ±0.05 ppm; this value should be used for conversion of data.
i Data from Ernst et al. [52]; this value should be used for conversion of data.
j Data from Engelhardt et al. [53], ±0.3 ppm.
k Data from Levy and Cargioli [20], ±0.05 ppm.
l Data from Lauterbur [17] for silicone "DC 200", used for reference signal in ref. [16,17].
m Data for $(CH_3)_3SiO[(CH_3)_2SiO]_xSi(CH_3)_3$ with $x = 50$ [50].
n Chemical shift given is for the atom with asterisk; reference compound suggested by Lauterbur [17].
o Data from Schraml [54] ±0.05 ppm.

ficient signal-to-noise ratio can be achieved in favorable cases with large n by routine measuring techniques [42,43], the recently developed INADEQUATE method holds great promise for such measurements, especially since isotopic enrichment is not likely to become common for silicon-29 compounds. INADEQUATE (incredible natural abundance double quantum transfer experiment) [44-46] achieves a good signal to noise ratio for satellites by a suitably chosen pulse sequence that among other things suppresses the main line.

Referencing and Solvent Effects. Fortunately, the δ scale is now generally accepted for reporting ^{29}Si chemical shifts. In this scale the chemical shifts are given in ppm units relative to the signal of $(CH_3)_4Si$ (TMS); positive values correspond to paramagnetic shifts (i.e., to less shielded nuclei with low-field or high-frequency shifts).

The measurements of absolute resonance frequencies and the merits of their use in chemical shift calculations have been discussed by Harris [5]. The method can give a better signal to noise ratio since it does not require the addition of TMS to the measured solution, but it is essentially a method of external refencing.

For practical reasons several secondary references are in use; the appropriate conversion factors are given in Table I. In converting chemical shifts from the literature, care should be taken to select conversion factors determined under comparable conditions. Extensive tables of coupling constants and of properly converted chemical shifts accompany the reviews of Marsmann [3] and of Williams and Cargioli [4]. The former lists the shifts according to their values within various classes of compounds; the latter review lists the compounds according to their summary formulas.

Internally referenced chemical shifts can be measured on current FT spectrometers with a precision better than ±0.05 ppm. This precision allows studies of small structural effects. It has already been mentioned that much of the data on carbon-functional compounds was obtained with a considerably lower precision (±0.3 ppm), which thus inhibits a detailed interpretation.

NMR SPECTROSCOPY

Solvent effects on ^{29}Si chemical shifts are often assumed to be small or negligible, but in fact very little is known, especially about carbon-functional compounds. The very few studies of solvent effects on silicon shielding included reference compounds (TMS [55,56], HMDSO [50], Si(OC$_2$H$_5$)$_4$ [50]), triphenylchlorosilane [57], and trimethoxysilane [57], in which the solvent effect did not exceed 0.7 ppm, several silylamines [57] with solvent effects less than 4 ppm, and silanols [57] with a variation of solvent shifts more than 5 ppm. The solvent effects in silatranes reach [15,58] 14 ppm.

3.2.2 ^{29}Si Chemical Shifts - Basic Facts

Before discussing the interpretation of the observed chemical shifts and the effects of a remote substituent (Y), some basic facts about the trends in silicon-29 shielding should be briefly reviewed.

The total range of ^{29}Si chemical shifts spans some 400 ppm; most of the known shifts are, however, clustered in a narrower range of about 200 ppm. Chemical shift ranges characteristic of various organosilicon compounds are shown in the chemical shift chart in Fig. 1.

Naturally, α substituents* (X) directly attached to silicon have a dominating influence on its shielding. The number and nature of X substituents determine in which spectral region the resonance of a particular silicon compound occurs. Within that region the chemical shift is modified by the more remote Y substituents that have only a secondary influence.

---

*It is customary in organic chemistry to designate derivatives as α, β, γ, etc., in dependence on the relative position (to the main functional group) of the carbon atom to which the second functional group is attached, e.g., (CH$_3$)$_3$SiCH(Cl)CH$_3$ is labeled as an α-chloro derivative since chlorine is bonded to the α (relative to the silicon atom) carbon. In NMR spectroscopy the observed effects are named in relation to the number of bonds separating the functional group from the observed nucleus. Thus, in the above example, silicon experiences a β-effect from the chlorine atom. The two naming systems should not be confused.

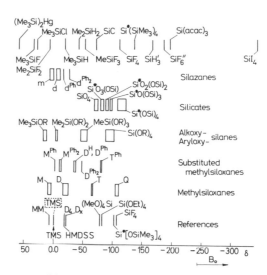

Fig 1. Chart of ^{29}Si chemical shifts. (Me = CH$_3$, Et = C$_2$H$_5$, Ph = C$_6$H$_5$, acac = acetylacetonate, R = alkyl or aryl, M^{X}i = Me$_{3-i}$X$_i$SiO, D^{X}i = Me$_{2-i}$X$_i$SiO$_2$ and T^{X}i = Me$_{1-i}$X$_i$SiO$_3$; for other symbols see Table I. Adapted from [6]).

The substituents influence the shielding through (a) polarization of the Si-X bonds, (b) back-bonding, (c) variation in bond angles and lengths and steric interaction, (d) electric field, (e) magnetic anisotropy, and (f) possibly other factors as well. Since these effects are not independent, the relative importance of each is difficult to assess with the exception of effects due to magnetic anisotropy of aromatic ring currents, which are considered to be small. Nevertheless, a consensus has been reached, and the polarization of Si-X bonds (or the total charge on the silicon atom) is generally accepted as the primary factor influencing the δ (Si) values. As will be discussed later, the effects of back-bonding are assumed to be of comparable magnitude (See section 3.2.3), but this assumption is not supported by any quantitative theory. The dependence of the silicon shielding on factors related to the polarity of the Si-X bonds is well documented in the literature.

Most of the reported ^{29}Si chemical shifts roughly fit the gross dependence on the electron density (Fig. 2). Depending upon the author's choice or data availability, the chemical shift is plotted against one of the following parameters: the number of electronegative substituents n [12,17,59], the sum of substituent electronegativities $\sum \chi$ [52], the sum of substituent constants $\sum \sigma$, or the net or total atomic charge Q_{Si} calculated by various quantum chemical methods [60,61]. Though the details of the dependence (curvature, slope, position of the minimum) are affected by the chosen parameter and the individual series of compounds, the "sagging" pattern of the dependence is a general one. The first electron-withdrawing substituent causes a decrease in the shielding (A-branch), and after some leveling off, additional substituents have the opposite effect (B-branch). Explanation of the "sagging" pattern has been the main objective of ^{29}Si shielding theory (See section 3.2.3).

The non-monotonous behavior of the ^{29}Si chemical shifts shown in Fig. 2 reduces somewhat the analytical potential of ^{29}Si-NMR for determination of the structure of unknown compounds. It is not possible to construct a simple and general direct additivity scheme for ^{29}Si chemical shifts, as is the case in, e.g., ^{13}C NMR spectroscopy, that would hold for the whole range of ^{29}Si chemical shifts. If direct additivity of α substituent effects is replaced by pairwise additivity [62], chemical shifts can be predicted with better accuracy (±6 or 11 ppm depending whether or not the -OR substituents are considered), but the accuracy is not sufficient for analytical purposes, and the number of required mutual interaction parameters increases rapidly with the number of substituents considered (for an extensive table of these parameters, see the review by Marsmann [3]).

It should be stressed that the "sagging" pattern describes only the main gross features of the effects of electron density on shielding. Since other factors also influence the observed shift, the dependence does not necessarily give correct predictions of trends in narrow ranges.

A study of ^{29}Si chemical shifts in carbon-functional compounds, $X^1X^2X^3Si(CH_p)_mY$, can be viewed as a probe to measure the slope of the general dependence at a par-

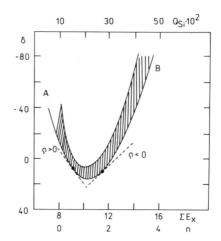

Fig. 2. General dependence of ^{29}Si chemical shifts, δ, on electron density. (The density is approximated by the total net charge (Q_{Si}), or by the number (n) of electronegative substituents, or by the sum of their electronegativities (E). Two selected points with positive (A-branch) and negative (B-branch) slopes (ρ) are shown).

ticular point. The substituents X^1, X^2, and X^3 determine at which point the dependence is probed (the A or the B branch, with positive and negative slopes ρ, respectively). The slope predicted according to the dependence on Fig. 2 can be compared with the observed effects of the Y substituents. Several examples have been described [13,53,63] in which the observed effects contradict the general trend. The examples involve transmission of substituent effects through the O-Si bond.

3.2.3 ^{29}Si Shielding Theory and Model

The chemical shifts $δ^A$ that is observed for nucleus A in a sample is equal* to the difference of the shielding

---

* Except for a numerical factor of 10^6, which has been introduced in order to maintain the above definition of the δ scale for the chemical shift, and which is ignored in the text.

constants σ^R and σ^A in the reference (R) and in the sample (A)

$$\delta^A = (\sigma^R - \sigma^A) \cdot 10^6 \tag{1}$$

Both σ^R and σ^A can be written as the sums of the shielding constants of the isolated molecule (σ_M) and the contributions originating outside of the molecule: the overall bulk susceptibility effect (σ_b), contributions from anisotropy in the molecular susceptibility (σ_a), from van der Waals forces between the molecules (σ_w), from the electric field of neighboring molecules (σ_e), and from specific interactions (σ_s). Thus,

$$\delta^A = [-\sigma_M^A + \sigma_M^R + \Delta(\sigma_b + \sigma_a + \sigma_w + \sigma_e + \sigma_s)] \cdot 10^6 \tag{2}$$

where the symbol Δ denotes the difference between the corresponding terms of the reference and the sample.

The Δ terms of Eq. (2) can be evaluated or eliminated through proper referencing, susceptibility corrections, and gas phase or solvent and concentration effect studies, but such studies have thus far been concerned very little with ^{29}Si NMR (see above), and there is no reliable estimate of the total magnitude of the Δ terms for silicon. Usually, it is estimated that in a series of related compounds the variation in the sum of the Δ terms does not exceed 1-2 ppm and that the shielding constant σ_M^A is identified with the chemical shift.

Quantum mechanical calculations of the shielding constant σ_M^A of a nucleus in an isolated molecule require the use of second-order perturbation theory. Even in its most general form [64] the Ramsey equation for the shielding (after proper averaging of the tensor components) can be written

$$\sigma_M^A = \sigma_d^A + \sigma_p^A \tag{3}$$

where σ_d^A and σ_p^A are the diamagnetic and paramagnetic terms, respectively. The diamagnetic terms depend only on the unperturbed ground state wave functions, while the paramagnetic ones depend on excited states as well.

Serious fundamental and practical problems are associated with the application of the formulas for $\sigma_p{}^A$ and $\sigma_d{}^A$, (See, e.g., refs. [64,65]): (a) too little is known about the wave functions describing the higher states, especially about the continuum states, (b) $\sigma_d{}^A$ and $\sigma_p{}^A$ depend upon the origins chosen for calculations and for the magnetic vector potential; and when incomplete basis sets must be used, their sum is also origin dependent, (c) with increasing size of the molecule both terms tend to become large and have opposite signs; hence, the result is subject to considerable error.

A large amount of work has been done to overcome these difficulties. The various approaches with which the problems are considered, both on ab initio and semiempirical levels, have been periodically reviewed [66].

With only two exceptions, which include ab initio calculations [67] of shielding in silane (SiH_4) and calculations of average excitation energy [68,69] (which plays an important role in the calculations of shielding constants), shieldings of silicon-29 in organosilicon compounds have been treated only by a very approximate semiempirical method that is briefly outlined here under the name of the ERW theory.

The ERW theory was first proposed by Engelhardt et al. [59], and it was subsequently further refined and developed in a series of papers by Radeglia and Wolff [60,69-73]. The following is an attempt to review the ERW theory in its present form. It should be noted that in its original form [59] (as well as in some of its early stages [60,69-71]) it neglected some terms that are now incorporated. Accordingly, any criticism (for a review see, e.g., [3]) should be considered in relation to the pertinent stage of the ERW theory development. Similarly, the theory proposed independently by Ernst et al. [52] can be viewed now as a special simple form of the ERW theory.

The main approximatons of the ERW theory are:

(a) the diamagnetic contribution, $\sigma_d{}^A$, is neglected and the variations in this quantity are not considered;

(b) ring currents and neighbor anisotropy effects are also neglected;

(c) the so-called average energy approximation is utilized;

(d) only the shielding contributions from 3s, 3p, and 3d orbitals centered on the silicon atom in question are considered;

(e) the radial parts of the p and d orbitals in the explicit expression for the shielding are approximated by Slater-type orbitals and are evaluated according to the Slater's rules;

(f) remote substituents are considered to affect the shielding by the electric field effect;

(g) and, finally, those approximations inherent in the semiempirical quantum chemical method employed in the calculation of molecular electron density distribution.

According to approximations (a), (b), and (f), the shielding of a ^{29}Si nucleus in an isolated molecule is the sum of two terms

$$\sigma^A = \sigma_p^A + \sigma_E^A \qquad (4)$$

where E denotes the minor contribution due to the electric field of remote substituents and p denotes the dominating paramagnetic contribution due to the restricted motion of electrons. The discussion will be begun with an evaluation of the σ_p^A term that is the central problem of the ERW theory, and then the role of σ_E^A term will be discussed.

With approximations (c) and (d) the paramagnetic shielding contribution can be expressed by the Jameson-Gutowsky formula [74]

$$\sigma_p^A = - \frac{C}{\Delta E_{av}} \left\{ \langle r_p^{-3} \rangle P_u^A + \langle r_d^{-3} \rangle D_u^A \right\} \qquad (5)$$

The positive constant C is the product of several universal physical constants, ΔE_{av} is the average excitation energy, and the average inverse cubes of the distances of the p and d valence electrons $\langle r_p^{-3}\rangle$ and $\langle r_d^{-3}\rangle$, are multiplied by P_u^A and D_u^A, respectively, which are complex expressions [74] representing the asymmetry of the corresponding electron populations.

Irrespective of the type of orbitals actually used in quantum chemical calculations leading to the values of the P and D terms, the radial terms are derived according to approximation (e) from Slater-type orbitals:

$$\langle r^{-3}\rangle = k\,(Z^*)^3 \tag{6}$$

Here $\underline{k}$ is a simple function of the principal quantum number and Z^* is the effective nuclear charge.

Extension of the validity of Slater's rules to fractional orbital populations gives the effective nuclear charges for 3s, 3p, and 3d electrons:

$$Z_p^* = Z_p^{*o} + 0.35(z_e - Q_{sp})f \tag{7a}$$

$$Z_d^* = Z_d^{*o} + (z_e - Q_{sp}) - 0.35 Q_d \tag{7b}$$

Symbol Z^{*o} denotes the effective nuclear charge for s, p, or d orbitals of a free uncharged silicon atom, z_e is the number of valence electrons, Q_{sp} is a sum of the traces of 3s and 3p submatrices of the electron density matrix, and Q_d is the trace of its 3d submatrix. The correction factor f is introduced in order to allow an empirical adjustment of the effect of the charge on the p-radial term (no such correction is applied to the d-radial terms). Since the value of f is optimalized for the best agreement between calculated and observed chemical shifts, its value also corrects for the inadequacy of the average energy approximation [69] (even if different optimum values are chosen for different series of compounds [69,71]).

A somewhat different approach to Eq. (6) was adopted by Roeland et al. [75]. They vary Z^* for each compound until an internal consistency with the charge calculated according to the Slater's formula is achieved for the silicon atom. Their results [75] do not seem, however, to differ from those obtained by the authors of the ERW theory.

The values of Q_{sp} and Q_d as well as the values of P and D should be calculated by quantum chemical methods. Approximate formulas for EHT and CNDO/2 (with or without population analysis) methods are available [72].

The authors of the ERW theory [59,60,69-73] have chosen not to report the shielding constants σ but instead to give so-called relative shielding constants σ^* which are defined as

$$\sigma^* = \frac{\sigma}{\sigma^o} \qquad (8)$$

where σ^o is the shielding constant in a hypothetical molecule with unoccupied 3d orbitals and with nonpolar bonds. The relative shielding constant is always positive, and a larger value corresponds to a less shielded nucleus; it is related to the δ-scale values by the following equation [60,76]

$$\delta = k_1 + k_2(\sigma^*-1) \qquad (9)$$

where k_1 and k_2 are constants (the scaling constant k_2 is positive).

After substitution from Eq. (5) into Eq. (8), the relative shielding constant is

$$\sigma^* = R_p^* \frac{P_u}{P_u^o} + R_d^* \frac{D_u}{P_u^o} \qquad (10)$$

where

$$R^* = \frac{\langle r^{-3} \rangle}{\langle r^{-3} \rangle_p^o} \qquad (11)$$

with superscript o denoting the term for the hypothetical compound.

Results of (CNDO/2 and EHT) calculations, which include silicon 3d orbitals and the D part of Eq. (10), are conflicting in regard to the relevance of the d orbitals in such calculations. Values calculated by the CNDO/2 method agree with the experimental data most satisfactorily when the d orbitals of silicon are neglected [71,72]. Horn and Murrell [77] have pointed out, however, that as a result of parametrization difficulties, such calculations are often unreliable; some of the results are not substantiated by ab initio calculations [77,78]. In EHT calculations inclusion of these orbitals improves the agreement. In any case, the d orbitals play a secondary role in determining silicon shielding in the ERW theory. Thus, these more elaborate calculations provide justification for the success of the original crude form of the ERW theory [59,60], which gave explicit expressions (Eq. 12 and 13) for the dependence of the silicon shielding on the electronegativity of the substituents.

In its original form, the ERW theory [59] considered only the 3p orbitals of silicon, and the need to use quantum chemical calculations was obviated by the adaptation of a modified Letcher-Van Wazer [79] model of a tetracoordinated atom. After generalization [60] to tetrahedral $SiX^1X^2X^3X^4$ molecules, the theory gives the relative shielding constant as

$$\sigma^* = [1 + 0.0843f(4 - \sum_i h_i)]^3 \cdot [\frac{1}{2}\sum_i h_i - \frac{1}{6}\sum_{i<j} h_i h_j] \quad (12)$$

where h_i, the ionic character of the $Si-X^i$ bond, is defined as a function of the electronegativity of silicon, $\chi(Si)$, and of the first atom of substituent X^i, $\chi(X^i)$

$$h_i = 1.0 - 0.16[\chi(X^i)-\chi(Si)] + 0.035[\chi(X^i)-\chi(Si)]^2 \quad (13)$$

It is clear from Eq. (12) and (13) that with increasing substituent electronegativity (decreasing total charge on the silicon atom) the radial part R^* (i.e., the cubic term in Eq. 12) increases for positive values of the correction factor f, and the P part decreases. Interplay of these opposing trends is substantially affected by the value of the correction factor f.

With suitably chosen values of the scaling factor (k_2 of Eq. 9) and the correction factor, Eq. (12) imitates correctly the gross trend (Fig. 2 and 3) observed for the dependence of the silicon shielding on n or on the total

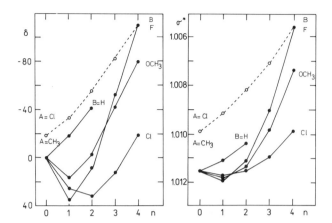

Fig. 3. Dependences of silicon-29 chemical shift (δ, part A) and relative shielding constant (σ^*, part B) on the number ($\underline{n}$) of substituents n compounds of the type $SiA_{4-n}B_n$ (Adapted from [59]).

charge on the silicon atom [59] in a number of compounds of the type $SiA_{4-n}B_n$.

Linear correlation between δ and σ^* (Eq. 9) is generally good for compounds of the $SiA_{4-n}B_n$ type [59,60]; deviations occur, however, either for compounds with multiple substitution by heavy atoms (Br, I) or for combinations of substituents of markedly different electronegativity. In the former case addition of an empirical term $k_3(\sum_i Z_i)^2$ to the right side of Eq. (9) improves the fit of the correlation (k_3 is an adjustable constant and Z_i is the atomic number of the i-th ligand). Deviations of the latter type can be made smaller if Eq. (12) is corrected for the non-tetrahedral arrangement of the substituents in molecules of the SiA_3B and SiA_2B_2 types. This correction can also be expressed as a function of substituent electronegativity with an empirical factor [70].

Attention will now be focused on the second term, σ_E^A, in Eq. (4). This term is particularly important for the discussion of silicon chemical shifts in carbon-functional compounds in which the shielding effects of the remote substituents (Y) are traditionally and conveniently discussed in terms of the substituent chemical shifts (SCS), $\Delta\delta$. Hence, the ERW theory uses the expression for SCS rather than a formula for σ_E.

The SCS of a Y substituent on the shielding of a nucleus is the difference between the (silicon) chemical shifts in the substituted compound (type B), $\delta(Si(CH_p)_mY)$, and in the parent compound (type A) $\delta(Si(CH_p)_mH)$. Thus,

$$\Delta\delta = [\delta(Si(CH_p)_mY)] - [\delta(Si(CH_p)_mH)] \quad (14)$$

According to the ERW theory [73] (approximation (f)), the effect of the remote substituent Y is essentially the effect of the electric field (as described by Buckingham [80]). The electric field is determined according to Coulomb's law by the distribution of net charges, which can be calculated by semiempirical quantum chemical methods. The formula for the SCS is

$$\Delta\delta = A[(a_1-E_H)\Delta\sigma^* + (a_2-\sigma_H^*)\Delta E] \quad (15)$$

where E_H and σ_H^* are the electric field and the relative shielding constants, respectively, calculated for the parent compound, and ΔE and $\delta\sigma^*$ are changes in these quantities due to the substitution of hydrogen by the substituent Y. The values of constants A, a_1, and a_2 can be determined by correlation analysis (together with the correction factor f); the values are constants for a given nucleus and the quantum chemical method used. This theory gives physical and quantitative justification to the suggestion [81] that the SCS is determined by a product of two factors, the first of which describes the sensitivity of the electron density to substitution, and the second measures the sensitivity of the shielding of the nucleus to the variation in the electron density. In Eq. (15) the quantities $\Delta\sigma^*$ and ΔE are determined by the substituent influence on the electron distribution in local and nonlocal terms, respectively, and the (a_1-E_H) and $(a_2-\sigma_H^*)$ factors describe the sensitivity of the shielding to the variation in electron distributions. With properly chosen values of all adjustable constants, Eq. (15) can account for the difference in SCS in substituted phenyl- and phenoxysilanes [73].

In chapter 4 of this volume, Ponec gives a current account of the theory of bonding in organosilicon compounds. He shows that the literature on this topic is plagued with misunderstandings and misconceptions. From the point of view of the ERW theory, the same can apparently be said about the interpretation of NMR data that employs concepts like back-donation, back-bonding, $(p-d)_\pi$ bonding, hyperconjugaton, etc. But since the model (as distinguished from the theory) based on these concepts gives useful qualitative predictions without recourse to elaborate calculations, we consider it to be of some heuristic value, especially for carbon-functional compounds, and it will be briefly described here.

The basic assumption of this model is that an increased electron density around the silicon nucleus brings about an increase in its shielding. The electron density detected in this way is a result of two (usually opposing) effects. The inductive effect is supposed to affect a silicon atom in the same way that it affects a carbon atom, i.e., the more electronegative the substituents bonded to the silicon atom, the more the electron density is reduced at the site of the silicon atom. Backbonding increases the electron density around silicon by interaction of unoccupied orbitals (silicon 3d orbitals or LUMO) with unshared electrons, i.e., $(p-d)_\pi$ interaction or bonding, or with π-electrons, i.e., $(\pi-d)_\pi$ interactions. The different upfield shifts are then interpreted as demonstrating the different abilities of various substituents to back-bond. With this model, interpretation of the "sagging pattern" in the dependence of the chemical shift on the number of substituents is straightforward. A very interesting situation occurs, however, in carbon-functional compounds with remote substituents Y that cannot enter into some special type of interaction (α-effect, hydrogen bond, etc.) and can act only through their inductive effects. Unless the substituent X directly bonded to the silicon atom is back-bonded, or unless the connecting carbon chain is back-bonded to the silicon, the effect on the silicon shielding should be a small paramagnetic shift for the electron-withdrawing substituent, Y. The paramagnetic shift would be larger if the connecting carbon chain is back-bonded to the silicon (as, e.g., in aromatic carbon-functional compounds), since the electron-withdrawing substituent weakens the back-bonding interaction. When the substituent X is back-bonded, the electron-withdrawing substituent Y enhances this bonding with the result that a diamagnetic shift is observed. A number of examples illustrating these generalized statements will be given in subsequent sections.

3.2.4 ^{29}Si Spin-Spin Coupling Constants

Spin-spin couplings of a silicon nucleus with nuclei such as ^{1}H and ^{19}F, the NMR spectra of which are measured with ease, have been frequently reported and studied. The results have been reviewed in considerable detail [3-6]. The more recent reviews also reflect the progress in NMR instrumentation that has made possible measurements of couplings with nuclei of lower receptivity (e.g., ^{29}Si-^{29}Si or ^{29}Si-^{13}C couplings).

The analytical potential of ^{29}Si coupling constants is rather limited. With the exception of one-bond couplings, the absolute values of the coupling constants are small, the values vary in narrow ranges, and their relation to molecular constitution is not always obvious. The direct coupling constants can be measured with good relative precision, their absolute values are large, and they are sensitive to the structure of the molecule [3-6]. Silicon coupling constants will be discussed in the sections on the appropriate carbon-functional compounds.

3.3 ALIPHATIC CARBON-FUNCTIONAL COMPOUNDS

The saturated methylene chain renders high conformational mobility to the carbon-functional compound. Since the motion around various C-C bonds is usually fast (on the NMR time scale), only the conformationally weighted average chemical shifts and coupling constants are observed, and the spectra are simplified. The simplification is a welcome feature if the spectra are recorded for analytical purposes. When, however, detailed interpretation of the spectra in terms of molecular geometry is attempted, the simplification means a loss of information. In addition, supporting quantum chemical calculations are complicated since they should be performed for all relevant conformations with optimized geometries in the rest of the molecule. Such a procedure increases the required computer size and time to the extent that sophisticated calculations (not only ab initio but even CNDO/2 approximations) for molecules of practical importance are out of the question and must be replaced by simpler semiempirical methods like the Del Re method.

For a specific chain length $\underline{m}$ and a specific functional group Y, some particular conformation might be energetically favored. When such a favored conformation leads to a distinct molecular property (e.g., reactivity, basicity, etc.) so that the compound stands out of a series of similar compounds, an explanation is usually sought in terms of a specific interaction. For example, conformational mobility can bring substituent Y to the proximity of the silicon atom, which allows a through-space interaction between Si and Y. A number of mechanisms for such interactions have been proposed in the literature (see also the other chapters). Several NMR studies were initiated with the hope of finding evidence for such specific interactions.

The interaction could be reflected in ^{29}Si NMR spectra, in the spectra of nuclei present in the functional group Y, or in the ^{13}C and ^{1}H NMR spectra of the connecting chain. Only ^{1}H NMR, however, has a real chance to elucidate the conformation of carbon-functional compounds in solution*, and for this reason the discussion will be begun with ^{1}H NMR spectroscopy of aliphatic carbon-functional compounds. Unfortunately, the current wave of interest in the NMR of "other" nuclei is at the expense of ^{1}H NMR work; hence, most of the available ^{1}H NMR data is of older vintage.

3.3.1 ^{1}H NMR Spectroscopy

The relatively high sensitivity of ^{1}H NMR makes it almost ideal for the rapid identification of compounds. The chemical shifts provide information on the type and

---

* The new technique INADEQUATE (Section 3.2.1) yields the values of ^{13}C-^{13}C coupling constants. Since these coupling constants are also conformationally dependent, they can be used for conformational studies. The INADEQUATE spectra of carbon-functional organosilicon compounds are strongly coupled and therefore their measurements and interpretation are difficult.

number of substituents in the methylene chain, and the splittings due to proton-proton couplings indicate structural relationships among various groups of protons. Characteristic values of proton chemical shifts and coupling constants will be given later for various classes of carbon-functional compounds. The small spectral dispersion of ^{1}H NMR frequently causes an overlap of lines and the spectra are of higher order. In such cases ^{1}H NMR spectroscopy does not give fast and unambiguous answers to analytical problems since the spectra must first be analyzed, which often requires considerable analyst and computer time. In such cases, it might be of advantage to use ^{13}C NMR spectroscopy, perhaps in combination with selective proton decoupling or two-dimensional NMR spectroscopy.

The most valuable contribution of ^{1}H NMR to the investigation of carbon-functional compounds with saturated connecting chains should be the determination of conformer populations or preferences and the energetics of conformational motion in solutions. Since this possibility is unique (other methods, though more precise, are applicable only to solids), and since conformation is so important to the chemistry of carbon-functional compounds (e.g., hyperconjugation and back-bonding have different conformational dependences) the potential of ^{1}H NMR should be examined in some detail.

The ^{1}H NMR methods that are available to serve this purpose are based either on the stereochemical dependence of coupling constants or on the dependence of direct dipole-dipole interaction on internuclear distances. The latter dependence correlates conformation with the results of measurements of the nuclear Overhauser effect, relaxation times, effects of shift and relaxation reagents, and with the spectra of oriented molecules. Of these methods, measurements of coupling constants are used most frequently.

Conformation and Coupling Constants. Practically all proton coupling constants (even direct one-bond couplings like $^1J(^{13}C-^1H)$ [82,83]) depend on molecular geometry. The effects of bond and dihedral angles and of other geometrical factors are described in the standard texts [84], and as this field is developing rather rapidly, new achievements are periodically reviewed [85]. The use of this method and its problems will be illustrated with vicinal or three-bond coupling constants, $^3J(^1H-^1H)$, which have been studied most intensively in the case of protons attached to sp^3-sp^3 carbon atoms (similar stereochemical dependences have been described, however, for protons on carbons of other types, e.g., on sp^2-sp^3 systems [86].

Theoretical calculations and ample experimental evidence (for reviews see [85,87-89]) have demonstrated that vicinal proton-proton coupling constants depend on the dihedral or torsion angle Φ between the two C-H bonds in a HC-CH fragment

At least the first four terms of the Fourier series expansion of the angular dependence

$$^3J(^1H-^1H) = A + B\cos\Phi + C\cos2\Phi + D\cos3\Phi + \cdots \quad (16)$$

are necessary [90-93] to describe adequately the dependence of the coupling constant on temperature and solvents. In practice, however, only the first three terms are retained in Eq. (16), which is then called the Karplus equation [94,95].

The simple picture expressed by Eq. (16) is in practice complicated by substituent effects that this equation does not take into account. Introduction of a substituent affects not only the dihedral angle of interest, but it also influences the coupling through the induced changes in hybridization and in bond angles and lengths [87]. These effects are not independent, and when two substituents are present, their effects are not additive. The combined effect of two substituents depends on their mutual orientation and on the orientation of the coupled protons.

A number of corrections to the Karplus equation have been proposed to account for the above mentioned effects [85,87-89,95]. Some of the corrections characterize the substituents by their general properties (e.g., electronegativity) [96]; some are expressed as functions of other measurable quantities (e.g., $^1J(^{13}C-^1H)$ coupling constants

[97], or chemical shifts of the coupled protons [98]). Corrections of the latter type are promising since the other quantities are also influenced in a complex way by the interrelated effects. Although some of such corrections have been subjected to strong criticism [99], others, based on arithmetic manipulations with two measurable coupling constants [100,101], are frequently employed for determination of conformations in cyclic compounds.

In practice, authors use (with all necessary precautions) the simplest electronegativity correction in the form [95,102]

$$^3J(^1H-^1H) = (A + B\cos\Phi + C\cos 2\Phi)(1-D\sum_i \chi(i)) \qquad (17)$$

(the summation goes over all substituents) and try to make a judicious choice of the correct values for the numerical factors.

In non-rigid systems the observed coupling constant, J_{obs}, is a motional average over all dihedral angle values, i.e.,

$$J_{obs} = \int_0^{2\pi} J(\Phi) \cdot P(\Phi) \cdot d\Phi \qquad (18)$$

where $P(\Phi) \cdot d\phi$ is the probability of the occurrence of the rotamer with the dihedral angle Φ. Calculations according to Eq. (18) require knowledge of the molecule's potential function for internal rotation [103]. Even for very simple potential functions the calculations lead to complex results [103,104]. Therefore, the simplest conformational motion is usually presumed.

Assuming that the internal rotation takes the form of jumping among the relatively stable rotamers with dihedral angles Φ_i, Eq. (18) simplifies to

$$J_{obs} = \sum_i J(\Phi_i) \cdot P(\Phi_i) \qquad (19)$$

where $P(\Phi_i)$ are the mole fractions of rotamers with the indicated dihedral angles. At equilibrium the ratios of

mole fractions are given by the difference in free
enthalpies of the corresponding rotamers

$$\frac{P(\Phi_1)}{P(\Phi_m)} = \exp\left(-[G(\Phi_1) - G(\Phi_m)]/RT\right) \qquad (20)$$

The standard procedure in the application of the
above relationships to the study of rotamer populations
starts with estimating $J(\Phi_i)$ values for all rotamers
according to one of the modifications of Karplus equation.
The derived values are inserted in Eq. (19) (taking proper
care for molecular symmetry) together with the experimentally determined value of J_{obs}. Solving this equation in
combination with the condition

$$\sum_i P(\Phi_i) = 1 \qquad (21)$$

gives the mole fractions of the rotamers which can give
the energetics of the considered motion by means of Eq.
(20).

This procedure can be illustrated on a rotamer
equilibrium in $CH_3SiH_2CH_2Y$ compounds (Y = Cl, I, $N(CH_3)_2$,
SCH_3, and H) studied by Carleer and Anteunis [105]. The
Newman projections of rotamers I, II, and III, which are
considered stable, are shown below with various couplings
indicated

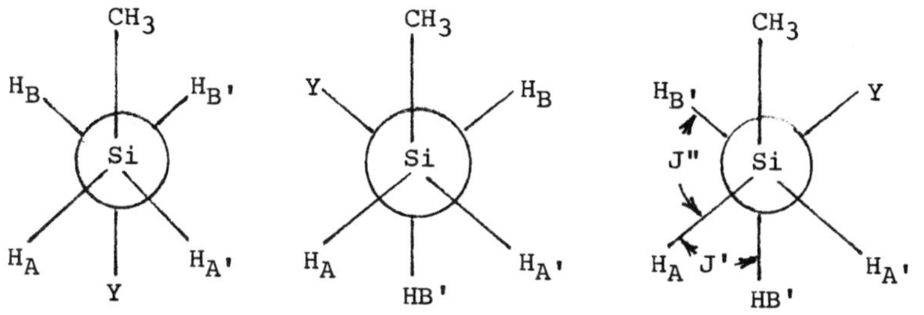

rotamer	I	II	III
mole fraction	P_I	P_{II}	P_{III}

Because of symmetry, rotamers II and III are equivalent ($P_{II} = P_{III}$), and Eqs. (21) and (20) take the forms

$$P_I + 2P_{II} = 1 \tag{22}$$

$$P_I/P_{II} = \exp[-(G_{II}-G_I)/RT] \tag{23}$$

The two vicinal coupling constants J_{AB} and $J_{AB'}$ which can, in general, be obtained from ^{1}H NMR spectra of the AA'BB' type are calculated according to Eq. (19).

$$J_{AB} = J(60°)_I P_I + J(180°)_{II} P_{II} + J(60°)'_{III} P_{II}$$

$$J_{AB'} = J(180°)_I P_I + J(60°)_{II} P_{II} + J(60°)''_{III} P_{II}$$

where the index relates the coupling constant to the rotamer, and the dihedral angle between the coupled protons is shown in parentheses. Carleer and Anteunis [105] assume

$$J(180°)_I = J(180°)_{II} = J_a$$

$$J(60°)_I = J(60°)_{II} = J(60°)'_{III} = J(60°)''_{III} = J_g$$

The observed spectra are the A_2B_2 type with only one observable vicinal coupling constant J that is the average of J_{AB} and $J_{AB'}$.

$$J = P_I(J_g + J_a)/2 + P_{II}(3J_g + J_a)/2 \tag{24}$$

In order to estimate the values of J_g and J_a coupling constants, the authors [105] use Eq. (17) with A = 4.15 (i.e., the value of ^{3}J(HSi-CH) in $H_3SiCH_2CH_3$), B = -A/8, and C = A-B. The electronegativity correction factors $D_\chi(Y)$ are calculated from the coupling in the corresponding H_3SiCH_2Y compounds for which a similar relationship holds:

$$^3J(H_3SiCH_2Y) = 4.56[1 - D_\chi(Y)] \tag{25}$$

The results, summarized in Table II, indicate that except for the dimethylamino derivative, the gauche rotamers (II and III) around the Si-C bond are more stable than the anti rotamer I. This is thought to be the result of two opposing effects: (a) increasing the C-Y bond length

Table II. Vicinal coupling constants and conformational analysis for $CH_3SiH_2CH_2Y$ compounds[a]

Y	$^3J(H_3SiCH_2X)$[b]	$^3J^c$	J_g^d	J_a^d	$2p_{II}/p_I$[e]	ΔG^{307}[f]	ΔH^{307}[g]
Cl	3.50	3.05	1.19	7.18	3.11	-2.89	-1.13
I	4.00	3.55	1.37	8.31	3.10	-2.80	-1.05
$N(CH_3)_2$	3.65	3.38	1.24	7.49	1.70	-1.38	+0.42

[a] Data from Ref. [105].

[b] The experimental value used for calculations according to Eq. (25).

[c] The experimental value of $^3J(HSi-CH)$ coupling in $CH_3SiH_2CH_2Y$.

[d] Calculated from Eq. (17).

[e] Molar ratio from Eq. (24).

[f] Calculated from the equilibrium constant $K = 2p_{II}/p_I$.

[g] Calculated as $\Delta H = \Delta G + RT \ln 2$, where the last term is due to entropy of mixing. Negative value indicates that gauche conformer is in excess.

reduces van der Waals repulsion, and bond angle deformations become more probable, and (b) increasing the van der Waals radius increases non-bonding repulsions [105-107].

The above discussion and example have clearly demonstrated on the one hand that the use of vicinal (or other) coupling constants for determination of conformer populations is straightforward and does not require any special sample preparation or measuring technique. It is easy to apply to new systems and to studies of temperature or solvent effects. On the other hand, because of the many approximations and assumptions necessary, the numerical values of the derived parameters are not very precise and reliable, though the qualitative conclusions about conformer preferences do appear sound. When a more detailed discussion of conformer populations is attempted, the results should be confirmed either (1) by measurements and analyses of other coupling constants with known stereochemical dependence (e.g., $^3J(^{19}F-^1H)$ [108], $^3J(^{13}C-^1H)$ [109], $^3J(^{13}C-^{13}C)$ [110], $^4J(^1H-^1H)$ [111,112], etc.) or (2) by other available methods based on direct dipole-dipole interaction.

Conformations and the Nuclear Overhauser Effect.
Measurements of proton-proton nuclear Overhauser effects (NOE) [7,113] are especially valuable, but they require careful preparation and degassing of the sample and special measurements. The measurement of NOE, $f_A(B)$, consists of measurements of intensities of NMR lines of the A protons with (I_{irr}^A) and without (I_o^A), saturation of the resonance of the B protons. The NOE is then defined as

$$f_A(B) = (I_{irr}^A - I_o^A)/I_o^A \qquad (26)$$

The measured NOE is related to various relaxation processes that provide polarization transfer from the B protons (saturated) to the observed A protons:

$$f_A(B) = \sum_B \frac{\gamma_B}{2 \cdot \gamma_A} \cdot \frac{\rho_{AB}}{R_A} - \sum_C \frac{\gamma_C}{2 \cdot \gamma_A} \cdot \frac{\rho_{AC}}{R_A} f_C(B) \qquad (27)$$

where γ_A and γ_B are magnetogyric ratios of protons A and B, R_A is the total relaxation rate of the A protons,

NMR SPECTROSCOPY 151

ρ_{AC} is the relaxation rate of the A protons due to dipole-dipole interaction of A protons with C protons, and the summation includes all B protons and all of the remaining C protons in the sample. The information on molecular geometry is contained in the ρ_{AC} terms, which depend on the internuclear distance (to the -6 power) between protons A and C. The formulas for NOE have been calculated for various spin systems [7,113] and for several types of internal motion [114].

In rigid systems NOE measurements give reliable internuclear distances. When the distances vary because of internal motion, the derived distances depend to some extent on the model of internal motion chosen for the calculation. In favorable cases the validity of the chosen model can be verified from NOE measurements on other pairs of protons within the same molecule.

<u>Conformation and Shift Reagents</u>. Similar problems in the interpretation of experimental results accompany the other methods that evaluate the internuclear distances from dipole-dipole interactions measured indirectly (through relaxation times or the effects of shift or relaxation reagents) [115]. The use of shift or relaxation reagents is also complicated by the complexing of the molecule under study with the reagent. Complexing affects the conformational equilibrium not only by the steric effect of the bulky ligands, but the coordination also alters the electronic distribution. Application of these techniques to various systems are well described in the literature [116]. Yasterbov et al. [117] determined the prevailing conformation in complexes of shift reagents with $(CH_3O)(CH_3)_2SiCH_2Cl$.

<u>Conformations and Oriented Molecules</u>. The interpretation of experimental results in terms of internuclear distances would be simplified if indirect evaluation of dipole-dipole interaction from measurements of NOE, relaxation times, and other parameters could be replaced by direct measurement of these interactions. In isotropic media the dipole-dipole interaction cannot be observed since the molecular motion averages out the components of the corresponding traceless tensor. In fully anisotropic situations (i.e., in crystalline solids) where inter- and intra-molecular proton-proton distances are comparable, the dipolar inter- and intra-molecular interactions have a

profound effect on the spectra. In addition to the
appearance of the dipolar coupling, the lines are
broadened and the dipolar interactions contain the intermolecular
contributions. Also, the conformations and
geometry in the solid state are different from those in
solution. The required intermediate situation with a partial
orientation of molecules and with limited motion,
which will average out intermolecular contributions but
retain intramolecular ones, occurs when the molecules
being studied are dissolved in liquid crystals.

Investigation of a compound by NMR spectroscopy in a
liquid crystalline solution (LX NMR) requires the selection
of a convenient solvent and measurement at a specified
temperature when the solvent is in a liquid
crystalline state. Any temperature gradient must be
minimized, and the spinning rate of the sample must be low
(unless the spinning axis is parallel to the magnetic
field). On some spectrometers precautions must be taken
to ensure a proper (linear) recording of spectra that
usually have a large spectral width. The recorded spectra
are analyzed in a fashion similar to ordinary NMR spectra,
except that an additional term describing dipolar interaction
must be included in the Hamiltonian. The values of
direct coupling D_{ij} that are extracted by the analysis
from LX NMR spectra can be related through known equations
to the average inverse cube of the internuclear distances.

The interested reader can find a detailed description
of the method with necessary references in one of several
existing reviews [118-120]. The goal of this discussion
has been to draw the attention of organosilicon chemists
to this method, which has not yet been applied to carbon-functional
compounds, although several simple organosilicon
compounds have been investigated [121-126].

To summarize, the potential of ^{1}H NMR spectroscopy to
determine the conformation of carbon-functional compounds
has thus far been exploited only to a very slight extent,
and only the most approximative method has been employed.
The relatively new methods that are based on dipolar
interaction hold promise for future investigation into
this important problem.

The use of ^{1}H chemical shifts for evaluation of
electron distribution and substituent effects is hampered

by a small chemical shift range, by relatively large
magnetic anisotropy and solvent effects, and, in some
instances, also by concentration dependence. The coupling
constant situation is similar. Most proton-proton
couplings are influenced by molecular geometry (vide
infra). Nevertheless, several empirical correlations have
been described for proton chemical shifts of carbon-
functional compounds. For example, methyl protons in
$CH_3SiX^1X^2X^3$ compounds correlate linearly with the sum of
Taft polar constants, $\sum \sigma^*$, for substituent X [128]. A
similar correlation with $\sum \sigma^*$ holds for other protons in
alkyl chains if the observed shifts are corrected for
magnetic anisotropy effects [129]. The chemical shifts
have also been correlated with electron charge densities
on the hydrogen atom in question [130] or with its linear
combination with the densities on adjacent carbon atoms
[131]. Valuable informaton about s-character in C-H bonds
can be derived form $^1J(^{13}C-^1H)$ couplings (see review
[127]).

Nevertheless, carbon chemical shifts appear better
suited for electron density evaluation.

3.3.2 ^{13}C NMR Spectroscopy

Two particular features of ^{13}C chemical shifts are of
great value for studies of aliphatic carbon-functional
compounds. These features are the direct additivity of
substituent effects and the linear relation to electronic
density. These two general characteristics are well
treated in several books [132-134] and review articles
[135].

For obvious reasons the electron distributions in
aliphatic carbon-functional compounds have been calculated
for correlation with chemical shifts only by the simplest
quantum chemical methods: the Del Re [130] and ω-HMO
[136] methods. The former method has been extensively
used in this laboratory on various carbon-functional compounds. Fig. 4 illustrates the quality of the linear
correlation between the net σ-electron charge Q_C on the
carbon atom and its chemical shift. The plot includes
results obtained for various functional groups Y (Y = H, Cl,
F, NH_2, OR, PR_2, etc.), and several different substituents
X on the silicon atom (X = H, F, OR, R, etc.). The solid

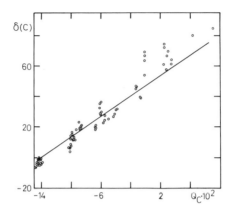

Fig. 4. Carbon chemical shifts in carbon-functional compounds vs. carbon total net charge calculated by the Del Re method. The solid line is the general dependence formulated by Lazzeretti and Taddei [30].

line is the original [130] least squares fit of the shifts in different types of compounds. The correlation holds irrespective of the nature and number of substituents attached to the carbon atom in question. The quality of the correlation permits its use for spectral assignments. Since the Del Re method considers only inductive effects, the other interactions (e.g., steric) that might be present should cause changes in the electron distribution. The altered distribution will be reflected in shifts that will deviate considerably from the correlation line.

Applicability of the direct additivity rule is also of some practical importance. The limited number of available data (and their low precision) does not permit a least squares treatment that would yield the "best" values of shielding contributions of various groups in the way that it does for other classes of organic compounds [132-134]. Good predictions are, however, obtained if the substituent effects of functional group Y are combined with the chemical shifts in the parent alkylsilane, $X^1X^2X^3Si(CH_2)_mH$ [137], or if the ^{13}C SCS values of the $SiX^1X^2X^3$ silyl groups of Table III are combined with the chemical shifts of the alkyl derivatives, $Y(CH_2)_mH$. As is apparent from Table III, the SCS values of the first two more diamagnetic with increasing chain length). Similar trends were also found for SCS values of other substituents [137].

An interesting correlation has been found [138] between ^{13}C chemical shifts and $^1J(^{13}C-^1H)$ coupling constants in carbon-functional compounds.

3.3.3 ^{29}Si NMR Spectroscopy

^{29}Si NMR Spectroscopy. The dominant influence of the α-substituents X on the silicon shielding in aliphatic carbon-functional compounds is apparent from Fig. 5. The shifts in the carbon-functional compounds follow the general pattern described earlier (Fig. 2). One of the consequences of the dominating role of the α-substituents is a linear correlation between the ^{29}Si chemical shifts in the $(CH_3)_{3-n}X_nSiCH_2Y$ carbon-functional compounds and in the $(CH_3)_{3-n}X_nSiCH_3$ parent compounds [6,139].

The complete ERW theory (i.e., with inclusion of the σ_E term) has not yet been applied to aliphatic carbon-functional compounds. Evaluation of the σ_E term would require good knowledge of the molecular geometry and the conformation. Without this term (which was shown to be important for aromatic carbon-functional compounds [73]), the ERW theory considers only the effects of remote substituents through their effects on electron distribution [140] as reflected in the relative shielding constant σ^*. The necessary quantum chemical calculations should, however, be carried out for optimized geometries.

Obviously, the effects of remote substituents on the electronic distribution at the site of the silicon atom are of a secondary order compared with the effects of α-substituents. This is confirmed by simple Del Re calculations and by the crude form of the ERW theory using Eq. (12) [140] that do not require optimization of molecular geometry. The results fit the general dependence of Fig. 2, but the overall quality of this gross dependence does not permit any discusson of possible deviations.

Some insight may be gained through the use of the substituent chemical shifts (SCS) $\Delta\delta$ defined by Eq. (14). (The ^{29}Si SCS value is the vertical distance in ppm between the corresponding points on the two lines in plots analogous to those in Fig. 5).

Table III. ^{13}C-Substituent chemical shifts[a]

Substituent R	m[b]	Effects[c]		
		α	β	γ (δ)
$(CH_3)_3Si$	1	2.1		
	2	1.9	0.3	
	3	4.0	1.5	2.6
	4	3.1	1.0	1.0 (0.2)
	5	2.4	0.4	0.8 (-0.8)
	5[d]	2.6	1.5	1.1 (0.1)
$(CH_3)_2(C_2H_5O)Si$	1	2.1		
	2	2.3	0.5	
	3	3.7	0.8	2.4
	4	2.4	-0.1	0.6 (-0.4)
$CH_3(C_2H_5O)_2Si$	1	-1.4		
	2	-0.2	0.4	
	3	0.9	0.4	2.2
	4	-0.3	-0.5	0.4 (-0.3)
$(C_2H_5O)_3Si$	1	-5.3		
	2	-3.4	0.2	
	3	-2.4	0.5	1.8
	4	-3.7	-0.7	0.0 (-0.8)

Table III - continued

Substituent R	m^b	Effects[c]		
		α	β	γ
$(C_2H_5)_3Si$	1	-4.9		
	2	-3.1	1.0	
$(CH_3)_2ClSi$	1	5.3		
	2	5.2	0.6	
CH_3Cl_2Si	1	8.9		
	2	8.9	0.8	
Cl_3Si	1	11.8		
	2	12.0	0.8	
	3	11.7	0.7	1.5

[a] SCS values in ppm calculated as SCS = $\delta(R-(CH_2)_mH) - \delta(H-(CH_2)_mH)$; positive values indicate paramagetic shift due to the substituent R. Data from Ref. [137].

[b] m is the alkyl chain length $H-(CH_2)_mH$ or $R-(CH_2)_mH$.

[c] The effects (SCS) are labelled according to the carbon's position relative to the substituent R.

[d] Data for the isomeric compound of the structure $(CH_3)_3SiCH_2CH_2CH(CH_3)_2$.

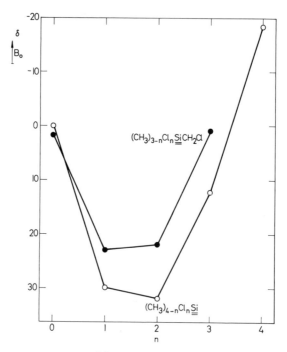

Fig. 5. Comparison of ^{29}Si chemical shift dependence on the number of α substituents in series with and without β chlorine atoms (Adapted from [6]).

In contrast to ^{13}C SCS values the ^{29}Si SCS values of a given remote substituent Y in a fixed position relative to the silicon atom are not constant (e.g., the SCS of a β-chlorine varies between +5 and -20 ppm) and their absolute magnitude decreases rapidly with increasing length of the connecting chain length $\underline{m}$. In general, the electronegative substituent Y (inductive donors have not yet been studied) appears to increase the silicon shielding more for electronegative substituents X, for a higher number $\underline{n}$ of substituents, and for a shorter connecting chain length $\underline{m}$. The dependence of ^{29}Si SCS values on the number $\underline{n}$ of substituents is illustrated in Fig. 6. The dependence [141] is approximately linear, and the slopes seem proportional to $\sigma_I$aliph of substituents Y.

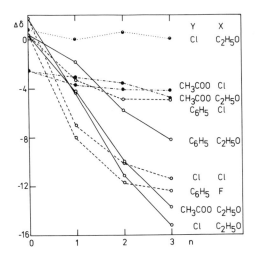

Fig. 6. ^{29}Si substituent chemical shift ($\Delta\delta$) dependence on the number and nature of X substituents for different Y functional groups in $(CH_3)_{3-n}X_nSi(CH_2)_mY$ compounds. (Open points refer to series with m = 1, full points to series with m = 2, and halved points to series with m = 3).

For all combinations of substituents X and Y* investigated thus far (X = OR, Cl, F, OC(O)CH$_3$; Y = R, OR, Cl, F, NH$_2$, P(C$_6$H$_5$)$_3$), the ^{29}Si SCS values become more negative with an increasing number $\underline{n}$ of substituents X that are electronegative, polarizable, and capable of backbonding. (That is, Y substituents cause increasing upfield shifts with increasing $\underline{n}$). This trend is in qualitative agreement when $\underline{n}$ > 0 both with the back-bonding model and with the ERW theory, since the values of $\underline{n}$ > 0 place the point corresponding to the parent compound (type A) on the B-branch of the dependence in Fig. 2 where the slope ρ is negative. The qualitative model can also accomodate better the stronger dependence of the SCS values on $\underline{n}$ when X = F than when X = Cl, although fluorine is less

---

* A classification of the SCS values published earlier [6] was derived on a basis of a smaller set of experimental results. Additional measurements (for Y = OR and F) did not confirm this classification for the case of $\underline{n}$ = 0.

polarizable and is a better $(p-d)_\pi$ donor than chlorine [142]. Points corresponding to $\underline{n} = 0$ demonstrate, however, the inadequacy of both approaches to the interpretation of silicon shielding. In $(CH_3)_3SiCH_2Y$ compounds (which are on the A-branch of the dependence in Fig. 2) Y substituents that are more electronegative than hydrogen should cause a decrease in the shielding relative to the parent $(CH_3)_3SiCH_3$, i.e., the SCS values of such substituents should be positive. The values are indeed positive for substituents such as CH_3 and C_2H_5, the inductive effects of which increase the total net (positive) charge on the silicon atom Q_{Si}. More electronegative substituents (Y = F, OH, and OR) cause a decrease of Q_{Si} (Del Re) by polarization of H_3C-Si bonds and thus lead to negative SCS values. However, when Y = Cl or when Y = NH_2, the Q_{Si} is also decreased, and yet positive SCS values are found (open points in Fig. 7). The simplest form of the ERW theory (Eq. 12) gives correct SCS values (i.e., $\Delta\delta^*$) for the series of $(CH_3)_{3-n}Cl_nSiCH_2Cl$ compounds [140]. Other carbon-functional compounds have not been subjected to this treatment, but the SCS values of twenty compounds with $\underline{m} = 1$ and Y = Cl correlate linearly with the Q(Si) charges calculated from the ionic characters h_i [140]. It should be noted that the discussed SCS

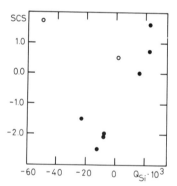

Fig. 7. Plot of ^{29}Si SCS ($\Delta\delta$) values against the total net charge on the silicon atom (Q_{Si}) in $(CH_3)_3SiCH_2Y$ compounds. (Open points correspond to compounds with Y = Cl or NH_2; these compounds have positive SCS values and Q_{Si} smaller than the reference compound).

values for n = 0 are small, and the corresponding points are close to the minimum of the dependence in Fig. 2 [140]. The discussed discrepancies suggest that not all factors have been taken into consideration. The direction of the deviations does not indicate that the dative Cl→Si or N→Si interactions are their cause.

3.3.4 Typical Results

Group I-Group IV Derivatives. NMR spectra of lithioalkylsilanes are influenced by aggregation and by lithium exchange. These processes are conveniently studied by ^1H and especially by ^7Li NMR spectroscopy. For example, lithiomethyltrimethylsilane was found to be tetrameric in hydrocarbon solutions [143], and measurements of ^7Li spin-lattice relaxation suggested that the tetramer is distorted away from a tetrahedral structure [144]. In mixtures with t-butyllithium an equilibrium is reached that involves random distribution of the two kinds of alkyl groups among tetrameric species [143].

Data on ^1H NMR spectra of $(CH_3)_3SiCH_2HgR$ compounds (Table IV) indicate very small effects of R substituents on the chemical shifts of both methyl and methylene protons. A change of solvent produces appreciably larger effects than the substitution. A noteworthy feature is the small absolute value of the $^2J(^{199}Hg-^1H)$ coupling constant in the symmetrical compound (R = $CH_2Si(CH_3)_3$). The $^1J(^{199}Hg-^{13}C)$ coupling constants in R_2Hg and RHgCl compounds are considerably smaller in trimethylsilylmethyl than in neopentyl derivatives [146]. The interpretation of this observation is analogous to that offered for $^1J(^{207}Tl-^{13}C)$ coupling later in this section.

Compounds having boron as the key atom in the functional group have often been investigated by ^{11}B NMR. Except for a few measurements of simple compounds (e.g., derivatives of borinic acid [147] like $((CH_3)_3Si)_2CH_2BOH)$, specialized NMR studies have considered almost exclusively compounds derived from higher boranes (e.g., [148]) or carboranes (e.g., [149]). Such studies have focused on problems of borane chemistry and their results have been recently reviewed [150].

The data in Tables V and VI provide information about

Table IV. ^{1}H – NMR data on $(CH_3)_3SiCH_2HgR$ compounds[a]

R	Solvent[b]	δ(CH$_3$)	δ(CH$_2$)	$^2J(^{199}Hg-^1H)$	$^4J(^{199}Hg-^1H)$
$(CH_3)_3SiCH_2$	neat	0.05	0.10	127.3	c
	CDCl$_3$	0.10	0.15	130.4	c
	C$_6$H$_6$	0.20	0.10	128.1	c
Cl	CDCl$_3$	0.11	1.03	249.9	7.1
	C$_6$H$_6$	-0.05	0.33	250.0	7.0
Br	CDCl$_3$	0.08	1.08	246.0	7.2
	C$_6$H$_6$	-0.11	0.44	252.8	6.7
I	CDCl$_3$	0.10	1.19	c	c
	C$_6$H$_6$	0.00	0.64	c	c

[a] Values converted from Ref. [145]. [b] Approximately 5-10% V/V solutions. [c] Not measured.

Table V. ^{1}H - NMR chemical shifts in some aluminum derivatives[a]

Compound	Conc.[b]	$\delta(CH_2Al)$	$\delta(CH_3Si)$	$\delta(CH_2O)$	$\delta(CH_3)C$
[(CH$_3$)$_3$SiCH$_2$]$_3$Al·O(C$_2$H$_5$)$_2$	100	-0.96	0.04	4.06	1.37
	77	-0.85	0.05	4.06	1.37
	53	-0.69	0.14	4.18	1.45
[(CH$_3$)$_3$Si(CH$_2$)$_3$]$_3$Al·O(C$_2$H$_5$)$_2$	100	—	0.08	4.09	1.41
(CH$_3$)$_3$SiCH$_2$AlCl$_2$	<10	-0.24	0.19	4.57	1.61
(CH$_3$)$_3$Si(CH$_2$)$_3$AlCl$_2$	—	—	0.08	—	—
[(n-C$_4$H$_9$)(CH$_3$)$_2$SiCH$_2$]$_3$Al·O(C$_2$H$_5$)$_2$	48	-0.84	0.08	4.20	1.46
[(n-C$_8$H$_{17}$)(CH$_3$)$_2$SiCH$_2$]$_3$Al·O(C$_2$H$_5$)$_2$	92	-0.92	0.09	4.00	—
(n-C$_4$H$_9$)(CH$_3$)$_2$SiCH$_2$AlCl$_2$	25	-0.10	0.22	4.61	—
(n-C$_8$H$_{17}$)(CH$_3$)$_2$SiCH$_2$AlCl$_2$	25	0.07	0.21	4.66	—

[a] Data converted from Ref. [151]. [b] Concentration of ethanol.

Table VI. NMR data on some thallium derivatives[a]

Compound[b]	CH$_2$ group				CH$_3$ group			
	δ(C)	δ(H)	^1J(Tl-C)	^2J(Tl-H)	δ(C)	δ(H)	^3J(Tl-C)	^4J(Tl-H)
R$_2$TlCl	30.6	1.27	1847	556,541[c]	5.5	0.16	168	18
R$_2$TlBr	–	–	–	536[c,d]	–	–	–	–
(CH$_3$)RTlBr	–	–	–	539[c,d]	–	–	–	–
R$_2$Tl-iBuAc[e]	28.7	1.07	1940	563,554[c]	1.8	0.17	168	18
(CH$_3$)RTl-iBuAc[e]	–	–	–	526[c,d]	–	–	–	–
RTl(-iBuAc)$_2$[e]	27.2	1.71	3540	1116	-1.1	0.17	318	33

[a] Unless indicated otherwise, data from Ref. [14]; measured in pyridine solutions. Tl stands for ^{205}Tl, C for ^{13}C, and H for ^{1}H.
[b] R stands for (CH$_3$)$_3$SiCH$_2$ group.
[c] Measured in CDCl$_3$ solution.
[d] Data from Refs. [152,153].
[e] iBuAc stands for OC(O)CH(CH$_3$)$_2$.

the ranges of NMR parameters in aluminum and thallium derivatives. The values are usually solvent and concentration dependent. Some trends have been noted. For example, similar values of $^2J(^{205}Tl-^1H)$ coupling constants in $((CH_3)_3SiCH_2)_2TlX$ and in $(CH_3)((CH_3)_3SiCH_2)TlX$ indicate that the CH_3 and $(CH_3)_3SiCH_2$ groups affect the s-character in the bonds between thallium and these groups similarly [152]. Smaller values of $^1J(^{205}Tl-^{13}C)$ couplings in R_2TlX than in $RTlX_2$ compounds are in agreement with a larger amount of thallium s-character in Tl-C bonds in the latter compounds [146]. This coupling in the neopentyl derivative is also larger than in the trimethylsilylmethyl derivative since the s-character of the carbon atom concentrates in bonds with less electronegative substituents [146].

The ^1H NMR spectra of carbon-functional compounds with $\underline{n}$ = 0 and Y = $Sn(CH_3)_3$, $Ge(CH_3)_3$, and $Pb(CH_3)_3$ were described in great detail by Schmidbaur [154] in 1964. Some of his results are summarized in Table VII. From a similarity with NMR parameters in $(CH_3)_4M$ compounds, he concluded that the substitution of a methyl proton by a trimethylsilyl group has a negligible effect on bonding within the $M(CH_3)_3$ groups. Significant changes occurred, of course, for the substituted CH_2 group. Interestingly enough, shifts of both signs were observed for the protons of the CH_2 group despite the electron donating properties of $M(CH_3)_3$ groups [154]. This fact only illustrates the difficulty mentioned earlier in interpretation of ^1H chemical shifts. The effects that substitution on silicon or germanium atoms have on the proton chemical shifts of compounds with the $SiCH_2Ge$ moiety are apparent from data in Table VIII.

An interesting case of diastereotopic nonequivalence of two $Si(CH_3)_3$ groups was noted [159] in a series of compounds with the structure $[((CH_3)_3Si)_2CH]_2SnXR$ (X = I; R = CH_3, C_2H_5, etc.). In consequence of the prochiral arrangement (for definitions of these terms see [160,161]) at the tin center of these molecules (which are monomeric in solutions), the two methyl proton lines are observed some 0.2 ppm apart (in C_6H_6 solutions). The signal separation can be reduced by addition of strong donor bases (dimethylsulfoxide, pyridine, etc.). The most plausible explanation assumes the formation of a stereochemically non-rigid, five-coordinate, donor-acceptor base adduct [159].

Table VII. ^{1}H - NMR parameters of $(CH_3)_3SiCH_2M(CH_3)_3$ compounds[a]

M	Si	Ge	Sn	Pb
$\delta(CH_3Si)$	0.03	0.01	0.00	0.01
$\delta(CH_3M)$	0.03	0.16	0.08	0.75
$\delta(CH_2)$	-0.26	-0.19	-0.26	+0.42
$^1J(^1H_3-^{13}CSi)$	118.2	118.0	118.2	119.0
$^1J(^1H_3-^{13}CM)$	118.2	124.4	127.4	136.0
$^1J(^1H_2-^{13}C)$	108.4	-	117.2	-
$^2J(^1H_3-^{29}Si)$	6.7	6.80	6.79	6.73
$^2J(^1H_2-^{29}Si)$	8.8	7.40	7.02	6.00

[a] Data taken from Ref. [154].

Group V Derivatives. A multinuclear approach to the investigation of carbon-functional compounds can be illustrated by a study of ^{13}C, ^{14}N, and ^{29}Si NMR spectra of (aminoalkyl)silanes [61], $(C_2H_5O)_n(CH_3)_{3-n}Si(CH_2)_mNH_2$, and (diphenylphosphinoalkyl)silanes [162,163], $(C_2H_5O)_n(CH_3)_{3-n}Si(CH_2)_mP(C_6H_5)_2$. Though the ^{1}H NMR spectra of amino derivatives were also measured [61] (Table IX), they were not utilized for conformer population analysis. The spectra indicated the existence of hydrogen bonding and reconfirmed the trends for ^{29}Si-^{1}H couplings established earlier. The proton chemical shifts could be related to the total net electron charges on the hydrogen atoms. Tables X and XI summarize the measured chemical shifts and the corresponding SCS values. The ^{29}Si and ^{13}C chemical shifts and SCS values, together with their partial interpretation, formed the basis of the earlier discussion in sections 3.3.2 and 3.3.3 and were utilized in the construction of Figs. 2, 4, and 6.

Table VIII. ^{1}H - NMR chemical shifts in some group IV derivatives

Compound	δ(CH$_3$)	δ(CH$_2$)
(CH$_3$)$_3$SiCH$_2$GeCl$_3$[a]	0.22	1.42
(CH$_3$)$_2$ClSiCH$_2$GeCl$_3$[a]	0.93	1.80
(CH$_3$)Cl$_2$SiCH$_2$GeCl$_3$[a]	0.83	1.66
[(CH$_3$)$_3$SiCH$_2$]$_2$GeCl$_2$[a]	0.16	0.85
[(CH$_3$)$_2$ClSiCH$_2$]$_2$GeCl$_2$[a]	0.60	1.33
[(CH$_3$)Cl$_2$SiCH$_2$]$_2$GeCl$_2$[a]	0.89	1.53
[Cl$_3$SiCH$_2$]$_2$GeCl$_2$[a]	-	1.73
[(Cl$_3$Si)$_2$CH]$_2$GeCl$_2$[b]	-	2.70[c]
(Cl$_3$Si)$_2$CHGeCl$_3$[b]	-	2.56[c]
[(CH$_3$)$_3$SiCH$_2$]$_2$Ge(CH$_3$)$_2$[a]	0.13(Si), 0.30(Ge)	-0.07
[(CH$_3$)$_3$SiCH$_2$]$_4$Pb[d]	0.05	0.50
[(CH$_3$)$_3$SiCH$_2$]$_3$PbCH$_3$[d]	0.05	0.45
[(CH$_3$)$_3$SiCH$_2$]$_3$PbCl[d]	0.15	0.25
[(CH$_3$)$_3$SiCH$_2$]$_4$Ti[e]	0.35	2.38
[(CH$_3$)$_3$SiCH$_2$]$_4$Zr[e]	0.28	1.18
[(CH$_3$)$_3$SiCH$_2$]$_4$Hg[e]	0.30	0.57
H$_3$GeCH$_2$SiH$_3$[f]	-	-0.01
H$_3$GeCH$_2$SiH$_2$Cl[f]	-	0.40
H$_3$GeCH$_2$SiHCl$_2$[f]	-	0.64

[a] Data from Ref. [155], measured in neat solutions.
[b] Measured in CCl$_4$ solution. [c] Chemical shift of CH proton.
[d] Data from Ref. [156]. [e] Data from Ref. [157]. [f] Data from Ref. [158].

Table IX. ^{1}H – NMR data on $(C_2H_5O)_n(CH_3)_{3-n}Si(CH_2)_mNH_2$ compounds[a]

		Chemical shifts δ						Coupling constants[b]		
m	n	CH$_3$Si	CH$_3$C	CH$_2$O	SiCH$_2$	CH$_2$N	NH$_2$[c]	$^2J(^{29}Si\text{-}CH_3)$	$^2J(^{29}Si\text{-}CH_2\text{-})$	$^1J(^{13}CH_3)$
1	0	0.19	–	–	2.11	2.11	0.87	6.5	5.3	117.9
1	1	0.09	1.15	3.65	2.14	2.14	0.97	6.3	5.5	118.8[d]
1	2	0.08	1.19	3.75	2.13	2.13	0.64	7.0	6.5	119.0[d]
1	3	–	1.20	3.81	2.15	2.15	0.68	–	7.8	–
2	0	0.00	–	–	0.69	2.72	0.99	6.5	d	117.7[e]
2	2	0.06	1.18	3.72	0.75	2.76	(1.79)	6.0	d	119.6[e]
2	3	–	1.20	3.77	0.76	2.74	(1.74)	–	d	–

[a] Data taken from Ref. [61]. Chemical shifts in 10% CCl$_4$ solutions, measured from cyclohexane δ = 1.44 ppm; maximum error ± 0.04 ppm.
[b] Coupling constants are measured in 33% CCl$_4$ solutions; ± 0.2 Hz.
[c] Values in parentheses are for 33% solutions.
[d] Could not be determined.
[e] Only one satelite line visible, ± 0.6 Hz.

Direct or through-space interaction between nitrogen and silicon atoms has been proposed for an explanation of some properties of the α-carbon-functional (aminoalkyl)silanes (see also chapters 2 and 4 of this volume). Since the interaction in question is of the donor (N) → acceptor(Si) type, it should lead to an increased shielding of ^{29}Si and decreased shielding of ^{14}N. (In silatranes, a similar transannular Si→N interaction shifts the ^{29}Si signal some 20 ppm diamagnetically relative to triethoxysilane [58]). The interaction is most likely to take place in the trimethylsilyl derivatives, but in such compounds the SCS values are positive. Negative ^{29}Si SCS values (which are much smaller than one would expect on the basis of the above comparison with silatranes) are found for compounds with two or three ethoxy groups bonded to the silicon atom. The silicon chemical shifts correlate linearly with the relative basicity of the amino group in the same compound (Fig. 8).

Within experimental error the number of ethoxy groups n on the silicon has no effect on the ^{14}N shielding, but the length of the connecting chain, m, has a decisive influence. The chemical shifts fall into three distinct groups depending on the value of m; values around -377 ppm

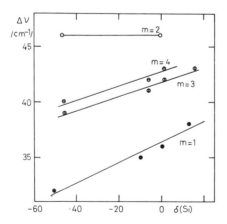

Fig. 8. The relative basicity (Δδ) of amino group vs. ^{29}Si chemical shifts in $(C_2H_5O)_n(CH_3)_{3-n}Si(CH_2)_mNH_2$ compounds.

Table X. ^{29}Si, ^{14}N, and ^{13}C chemical shifts in $(C_2H_5O)_n(CH_3)_{3-n}Si(CH_2)_mNH_2$ compounds[a]

		^{29}Si[b]		^{14}N[c]				^{13}C[b]		
m	n	δ(Si)	Δδ(Si)[d]	δ(N)	CH_3Si	CH_3C	CH_2O	$SiCH_2$	$CH_2(\beta)$[c]	$CH_2(\gamma)$[e]
1	0	0.5	+0.5	-376	-3.0	–	–	31.6	–	–
	1	13.2	-0.3	-376	-3.2	18.9	58.6	31.2	–	–
	2	-.9	-3.8	-378	-6.0	18.5	58.3	28.7	–	–
	3	-50.2	-5.7	-377	–	18.4	58.6	26.0	–	–
2	0	-0.2	-1.8	-348	-0.7	–	–	23.0	38.8	–
	3	-46.5	-0.6	-349	–	18.7	58.6	17.2	37.8	–
3	0	1.6	+0.9	-355	-1.1	–	–	14.4	29.3	46.4
	1	15.8	+1.0	-355	-1.6	19.2	58.5	14.2	28.5	46.2
	2	-5.7	+1.9	-355	-4.5	18.9	58.3	11.7	28.1	45.9
	3	-45.3	+1.7	-356	–	18.8	58.6	8.3	28.1	45.7
4	0	1.4	+0.8	-357						

Table X continued

^{29}Si, ^{14}N, and ^{13}C chemical shifts in $(C_2H_5O)_n(CH_3)_{3-n}Si(CH_2)_mNH_2$ compounds[a]

m	n	^{29}Si[b] $\delta(Si)$	$\Delta\delta(Si)$[d]	^{14}N[c] $\delta(N)$	CH_3Si	CH_3C	CH_2O	^{13}C[b] $SiCH_2$	$CH_2(\beta)$[e]	$CH_2(\gamma)$[e]
2		-5.9	+1.9							
3		-45.6	+1.8							

[a] Data from Ref. [61]. In ppm, positive values indicate low field shifts.
[b] TMS external reference, maximum error ± 0.3 ppm.
[c] CH_3NO_2 external reference, maximum error ± 3 ppm.
[d] Substituent chemical shift.
[e] Position relative to the silicon atom.

Table XI. ^{29}Si, ^{31}P, and ^{13}C Chemical shifts and ^{31}P coupling constants in $(C_2H_5O)_m(CH_3)_{3-n}Si(CH_2)_mP(C_6H_5)_2$ compounds[a]

Nucleus			^{29}Si			^{31}P		^{13}C
Group			$SiCH_2$			$P(C_6H_5)_2$		Si^*CH_3
m	n	J	δ	Δδ		δ	Δδ	δ
1	0	15.0	1.31	1.31		-30.7	-4	-0.2[b]
	1	17.09	15.26	1.76		-23.2	+3	-0.8[b]
	2	17.58	-9.70	-3.6		-23.6	+2	-3.8[c]
	3	15.14	-50.56	-6.06		-23.7	+2	-
2	0	20.89	2.54	0.9		-18.6	-6	-1.9
	3	31.25	-47.10	-1.2		(-9.1)	+3	-
3	0	0	1.13	0.4		-26.1	-9	-1.7
	2	0	-6.24	+1.36		-17.1	0	-2.74
	3	1.47	-46.50	+0.5		-17.3	0	-
4	0	0	1.46	0.9		-24.9	-8	-1.6
	3		-46.17	+1.2		-15.9	+1	-
5	3[d]	0	-45.73			-15.8		-
6	3[e]	0	-45.50			-16.0		-

Table XI continued

^{13}C							
Si*CH$_2$		SiC*CH$_2$		SiC$_2$*CH$_2$		SiC$_3$*CH$_2$	
J	δ	J	δ	J	δ	J	δ
29.3	14.5						
29.3	14.9						
29.3	12.7						
29.3	9.4						
11.0	12.0	14.6	21.6				
10.9	6.0	13.4	20.3				
11.0	18.4	17.1	20.3	13.5	32.1		
12.2	15.7	17.1	19.6	11.0	31.6		
12.2	12.1	18.3	19.4	11.0	31.3		
0	16.3	12.2	25.5	15.9	29.6	12.2	27.8
0	10.2	13.4	24.4	15.9	29.3	11.0	27.6
0	10.3	0	22.4	13.4	34.6	15.9	25.6
0	10.3	0	22.6	0	32.7	12.2	30.7[f]

[a] Data from Refs. [162,163]. The pertinent carbon is denoted by an upper left asterisk in the heading. The chemical shifts are in δ-scale, approximate error ± 0.05 ppm. ^{31}P shifts are relative to external 85% H_3PO_4. Coupling constants are in Hz, their signs were not determined, approximate errors ± 1 Hz.

[b] $^3J(^{13}C-^{31}P)$ = 3.7 Hz. [c] $^3J(^{13}C-^{31}P)$ = 2.4 Hz. [d] For SiC$_4$*CH$_2$ δ = 27.9. [e] For SiC$_4$*CH$_2$ δ = 24.3 and $^2J(^{13}C-^{31}P)$ = 15.9 Hz; for SiC$_5$*CH$_2$ δ = 28.0 and $J(^{13}C-^{31}P)$ = 11.0 Hz. The assignment of the two doublets is uncertain.

when $\underline{m}$ = 1, around -349 ppm when $\underline{m}$ = 2, and around -356 ppm if $\underline{m}$ ≥ 3. The ^{14}N SCS due to (ethoxymethyl)silyl groups are all positive and do not exceed 7 ppm. If the Si-N interaction could be compared with protonation, paramagnetic shifts of about 5-30 ppm would be expected for (aminoalkyl)silanes with such interactions. It is opposite to what is in fact observed.

The ^{13}C chemical shifts in (aminoalkyl)silanes fit the linear correlation with the net charges shown in Fig. 4. No deviations that would indicate N→Si interactions were noticed.

Also, no analogous interaction between Si and P could be detected in the ^{29}Si NMR spectra of (diphenylphosphinoalkyl)silanes. However, an ethoxy group bonded to the silicon atom had a marked effect on ^{31}P chemical shifts (Table XI), and in compounds with $\underline{m}$ = 1 the additivity rules for ^{13}C chemical shifts were violated. It was suggested that the two observations had a common origin in an interaction between the phosphorus and oxygen atoms. The interaction apparently did not involve the silicon atom.

The fact that the $^2J(^{31}P^{-1}H)$ coupling constant is 5-6 Hz larger in (trimethylsilylmethyl)phenylphosphonium salts than in the corresponding neopentyl compounds was interpreted [164] in terms of p_π-d_π hyperconjugation involving the α-hydrogen and the 3d orbitals of silicon and phosphorus:

$$P^+CH_2Si(CH_3)_3 I^- \longleftrightarrow \overset{H^+}{PCHSi(CH_3)_3 I^-}$$

This interpretation was also in agreement with other ^{31}P and ^{13}C NMR data.

Group VI Derivatives. (Acetoxyalkyl)silanes, $(C_2H_5O)_n(CH_3)_{3-n}SiCH_2OC(O)CH_3$, have been investigated [139] by NMR spectroscopy involving all NMR-active nuclei present in the molecules. The ^{17}O NMR results (Table XII) provided no new information. The shifts of oxygens in ethoxy groups varied in a fashion similar to those in methylethoxysilanes. The oxygens of the acetoxy groups have such wide lines that the chemical shifts of the C(O) oxygens could not be measured, and the linewidth of the C-O-C oxygen line is larger than the change in the chemical

Table XII. ^{17}O NMR data for $(C_2H_5O)_n(CH_3)_{(3-n)}SiCH_2OC(O)CH_3$ compounds[a]

n	$O(C_2H_5)$		$O(COCH_3)$	
	$\delta(^{17}O)$[b]	ω[c]	$\delta(^{17}O)$[d]	ω[c]
1	10	150	132	250
2	19	180	130	250
3	14	200	e	–

[a] Data from Ref. [139]. Chemical shifts in ppm units, relative to external H_2O reference. Positive values indicate shift to low field of reference line.

[b] Standard deviation ± 4 ppm.

[c] Line widths.

[d] Standard deviation ± 7 ppm.

[e] Too wide to be measured.

shift. For these reasons, the time-consuming measurements of ^{17}O NMR spectra were not made on other oxygen-containing carbon-functional compounds. Similarly, little information about the bonding situation could be derived from ^1H NMR spectra (Table XIII). Conformational analysis has not yet been carried out. ^1H NMR spectra were employed to elucidate the behavior of compounds in solutions and to assess the possible effect of solvent on other NMR parameters [139], e.g., the association of (hydroxyalkyl)silanes [165], and the acidity of silylcarbinols [166]. Dyer and Lee [167] determined a number of heteronuclear coupling constants from the ^1H NMR spectra of $(CH_3)_3Si(CH_2)_mOR$ (R = H, CH_3, $C(O)CH_3$, $Si(CH_3)_3$) and their thio analogs.

Some ^{29}Si and ^{13}C chemical shifts and SCS values are collected in Tables XIV and XV for $(CH_3)_{3-n}X_nSi(CH_2)_mOR$ compounds. The values (including some large values for

Table XIII. ^1H - NMR data $(CH_3)_{3-n}X_nSiCH_2OC(O)CH_3$ compounds[a]

X	n	$\delta^{b,c}$						
		CH_3-Si	CH_3-	$-CH_2-O$	$CH_3C(O)$	CH_2Si	$(i-)--CH_3Si---(j)$	$C_2H_5(k)$
C_2H_5O	0	0.027	–	–	1.992	3.708	6.6	119.5
		0.068	–	–	1.972	3.685	6.7	120.1
	1	0.131	1.138	3.668	1.998	3.730	6.8	119.59	6.9
		0.140	1.162	3.667	1.989	3.694	(6.8)	119.49	6.9
	2	0.149	1.163	3.760	2.002	3.707	7.1	120.19	6.9
		0.136	1.195	3.759	1.934	3.659		119.09	7.0
	3	–	1.182	3.823	2.012	3.746			7.2
		–	1.211	3.815	1.997	3.676			7.0

Table XIII continued

^1H - NMR data from $(CH_3)_{3-n}X_nSiCH_2OC(O)CH_3$ compounds[a]

δ[b,c]

X	n	CH_3-Si	CH_3-	$-CH_2-O$	$CH_3C(O)$	CH_2Si	$(i-)--CH_3Si---(j)$	$C_2H_5(k)$
$CH_3C(O)O$	1	0.282	1.990[h]		1.990	3.804	7.0	120.6
		0.308	2.003[d]		2.015	3.760	7.2	120.3
	2	0.498	2.017[h]		2.032	3.843	(8.2)	122.99
		0.496	2.036[h]		2.073	3.805		122.39
		–	2.058[h]		2.224	3.972	–	–

[a] Data of Ref. [139]. Chemical shifts in δ scale. Coupling constants J in Hz.
[b] Measured in 1:9 $CHCl_3$ solutions; where two values are given, that on the first line was obtained in $CHCl_3$ solution and that on the second line in CCl_4.
[c] Standard deviation ± 0.003 ppm
[d] The same as b; values in parentheses obtained in 2:1 solutions; maximum standard deviation ± 0.2 Hz.
[e] The same as b except that 2:1 solutions were measured; maximum standard deviation ± 0.2 Hz.
[f] Standard deviation ± 0.2 Hz.
[g] Only one satellite could be measured; estimated error ± 0.5 Hz.
[h] Acetoxy group of the substituent X.

Table XIV. ^{29}Si and ^{13}C-NMR chemical shifts in $(CH_3)_{3-n}X_nSi(CH_2)_mOC(O)CH_3$ compounds[a]

m	X^a	n	$^{29}Si^b$ δ	$δΔ°$	CH_3Si	$CH_2-α$	$CH_2-β$	$CH_2-γ$	$^{13}C^b$ C(O)	CH_3^d	OC^e	CH_3^e
1	OEt	0	0.3	0.3	−2.5	58.1			169.9	20.8	−	−
		1	9.0	−4.5	−3.4	56.2			170.1	19.8	58.3	18.2
		2	−16.1	−10.2	−5.5	53.6			169.9	19.8	58.2	18.1
		3	−58.2	−13.7	−	50.7			169.8	19.5	58.2	17.7
	OAc	1	13.8	−8.0	−3.0	56.5			171.0	19.5[f]	171.0	21.8[f]
		2	−18.1	−22.7	−2.2	55.5			173.6	19.1	170.2	21.9
		3	−82.5	−38.8	−	56.8			181.0	18.6	170.4	22.7
2	OEt	0	−1.0	−2.6	−2.6	16.6	60.8		168.4	19.6	−	−
		1	12.7	−3.1	−2.4	17.1	60.2		168.4	20.0	57.1	18.0
		2[g]	10.3[h]	−3.6	−5.1	14.8[h]	59.8[h]		168.7	19.7	57.0	17.6
		3	−50.7	−4.8	−	11.4	59.3		168.2	19.4	57.0	17.2

Table XIV continued

^{29}Si and ^{13}C-NMR chemical shifts in $(CH_3)_{3-n}X_nSi(CH_2)_mOC(O)CH_3$ compounds[a]

m	X^a	n	$^{29}Si^b$ δ	$\delta\Delta^c$	CH_3Si	CH_2-α	CH_2-β	CH_2-γ	$^{13}C^b$ C(O)	CH_3^d	OC^e	CH_3^e
3	Cl	1	28.2	-3.7	1.1	18.8	59.7		168.7	19.8	–	–
		2	29.9	-4.1	5.5	21.7	59.1		169.3	21.4	–	–
		3	10.4	-4.2	–	25.0	58.9		169.7	21.0	–	–
	OEt	0	0.9	0.2	-2.3	12.1	23.0	66.1	168.6	20.0	–	–
		2	-6.8	0.8	-5.0	10.3	22.7	66.3	169.6	20.5	58.0	18.5
		3	33.0	i	5.2	18.1	22.3	65.4	169.8	20.8	–	–

[a] Data from Ref. [139]. Abbreviations OEt = OC_2H_5 and OAc = $OC(O)CH_3$.
[b] Chemical shifts in δ-scale; error ± 0.3 ppm.
[c] Substituent chemical shift (SCS) of the acetoxy group.
[d] Methyl carbon of the acetoxy groups.
[e] Carbons of the X substituents.
[f] The assignments of the two lines can be interchanged.
[g] Because of sample size, possible error of all chemical shifts = ± 1 ppm.
[h] Impurity lines at δ(Si) = -9.4, δ(C) = 13.3 and 58.7.
[i] Data on the parent propane derivative not available.

Table XV. ^{29}Si- and ^{13}C-NMR chemical shifts in $(CH_3)_3Si(CH_2)_mOR$ compounds[a]

		^{29}Si			^{13}C					
m	R	δ	Δδ[b]	CH$_3$Si	C-α	C-β	C-γ	C-δ	(O-C)[d]	(CH$_3$)[c]
1	H	-2.5	-2.5	-4.4	54.2				–	–
	CH$_3$	-2.1	-2.1	-3.5	66.9				62.7	62.7
	C$_2$H$_5$	-2.6	-2.6	-3.9	63.6				69.8	14.4
	R[g]	-3.6	-3.6	-4.7	68.7				68.7	-4.7
	Si(CH$_3$)$_3$[h]	-2.0	-2.0	-4.6	53.8				69.9	-2.1
2	H	-2.0	-3.6	-2.6	20.4	57.3				
	CH$_3$	-0.1	-1.7	-1.4	18.1	69.4			57.3	57.3
	R[g]	-0.4	-2.0	-1.3	18.2	66.6			66.6	-1.3
	Si(CH$_3$)$_3$[i]	-1.3	-2.9	-1.9[f]	20.3	57.7			–	-2.4[f]
3	H	1.7	1.0	-1.3	12.8	27.4	64.9		–	–
	CH$_3$	0.4	-0.3	-2.6	12.2	23.5	74.8		57.2	57.2
	R[g]	0.0	-0.7	-2.4	12.1	23.5	72.5		72.5	-2.4

Table XV continued

^{29}Si- and ^{13}C-NMR chemical shifts in $(CH_3)_3Si(CH_2)_mOR$ compounds[a]

m	R	^{29}Si δ	Δδ[b]	CH$_3$Si	^{13}C C-α	C-β	C-γ	C-δ	(O-C)[d]	(CH$_3$)[c]
	R[g]	0.0	-0.7	-2.4	12.1	23.5	72.5		72.5	-2.4
4	H	0.4	-0.2	-1.9	16.2	19.8	36.1	61.0	–	–
	R[g]	0.4	-0.2							
	Si(CH$_3$)$_3$[j]	0.1	-0.5	-1.3[f]	15.6	19.4	35.7	60.8	–	-2.5[f]

[a] Data from Ref. [168]. All chemical shifts in δ-scale (ppm units, relative to external TMS, paramagnetic shifts positive); error ± 0.3 ppm.
[b] Substituent chemical shifts.
[c] The position of carbon atoms relative to the silicon atom.
[d] Structurally different carbon atoms.
[e] Methylcarbons of different types.
[f] The assignment of the two lines can be interchanged.
[g] R = $(CH_3)_3Si(CH_2)_m$.
[h] δ(^{29}SiC) = 16.1.
[i] δ(^{29}SiC) = 13.1.
[j] δ(^{29}SiO) = 13.9.

Table XVI. ^{13}C chemical shifts of methylene carbons calculated according to the additivity of SCS in $(CH_3)_3Si^*(CH_2)_mOR$ compounds[a]

Carbon[b]		C-α			C-β			C-γ			C-δ		
m	R	SCS[c]	δ[d]	Δ[e]	SCS[c]	δ[d]	Δ[e]	SCS[c]	δ[d]	Δ[e]	SCS[c]	δ[d]	Δ[e]
4	$(CH_3)_3Si^*$	3.1			1.0			1.0			0.2		
	H	13.9	17.0	-0.8	19.4	20.4	-0.6	35.3	36.3	-0.2	61.7	61.9	-0.9
	$Si(CH_3)_3$	12.9	16.0	-0.4	18.3	19.3	0.1	34.4	35.4	0.3	61.3	61.5	-0.7
3	$(CH_3)_3Si^*$	4.0			1.5			2.6			–		–
	H	10.3	14.3	-1.5	26.1	27.6	-0.2	63.9	66.5	-1.6			
2	$(CH_3)_3Si^*$	1.9			0.3			–		–	–		–
	H	17.9	19.8	0.6	57.3	57.6	-0.6	–					
	CH_3	14.4	16.3	1.8	67.4	67.7	1.7	–					
	$Si(CH_3)_3$	17.3	19.2	1.1	56.5	56.8	0.9	–					

Table XVI continued

^{13}C chemical shifts of methylene carbons calculated according to the additivity of SCS in $(CH_3)_3Si^*(CH_2)_mOR$ compounds[a]

		C-α			C-β			C-γ			C-δ		
m	R[b]	SCS[c]	δ[d]	Δ[e]	SCS[c]	δ[d]	Δ[e]	SCS[c]	δ[d]	Δ[e]	SCS[c]	δ[d]	Δ[e]
3	C(O)CH₃	13.5	15.4	1.2	59.5	59.8	7.0	—	—	—	—	—	—
1	(CH₃)₃Si*	2.1											
	H	49.3	51.4	2.8									
	CH₃	60.0	62.1	4.8									
	C₂H₅	57.3	59.4	2.1									
	C(O)CH₃	50.4	52.5	-5.5									

[a] Data from Ref. [168].
[b] Carbon atom position relative to the silicon atom with asterisk.
[c] SCS values of substituents R.
[d] The chemical shift calculated from the SCS values according to the additivity rule.
[e] Difference between the observed and calculated chemical shifts.

(acetoxymethyl) derivatives) confirm the general trends described earlier. Negative ^{29}Si SCS values in compounds with $\underline{n}$ = 0 were also discussed in Section 3.3.3. The small positive ^{29}Si SCS value observed in (acetoxymethyl)trimethylsilane is apparently due to the proximity of the carboxyl group to the silicon atom. The carboxyl group is known to have large electric field and magnetic anisotropy effects.

The ^{13}C chemical shifts follow the additivity rule in all cases when $\underline{m}$ ≥ 3. For shorter connecting chains sizable deviations occur (Table XVI), especially when R = C(O)CH$_3$. It is not clear whether the deviations are caused by altered conformation populations or by an interaction of the two substituents on the methylene chain. There is, however, no indication in support of the latter possibility.

Group VII Derivatives. (Haloalkyl) groups provide a convenient means for variation of inductive effects. For that reason many papers contain some NMR data on (haloalkyl)silanes, but only a few of them are aimed at furthering the understanding and knowledge of these carbon-functional compounds.

Detailed NMR investigations of (halomethyl)silanes, H$_3$SiCH$_2$Y, have been pioneered by Bellama and MacDiarmid [169]. Their results are summarized in Table XVII. Vicinal proton-proton coupling constants and their dependence on electronegativities as determined in this study [169] were utilized in the conformational analysis discussion in Section 3.3.1. The ^{1}H NMR results were later supplemented by ^{13}C and ^{29}Si NMR data on (chloromethyl)silane [170].

The ^{1}H NMR chemical shifts have been reported on virtually all possible (halogenoalkyl)silanes [171]. Some have also been investigated by ^{13}C and ^{29}Si NMR spectroscopy. Since the NMR spectra of the active isotopes of chlorine, bromine, and iodine are not yet routinely measured [172], only (fluoroalkyl)silanes have been investigated by NMR spectroscopy of all of the NMR-active nuclei present (in some instances NMR investigations of (chloroalkyl)silanes have been accompanied by ^{35}Cl and ^{37}Cl NQR data, e.g., [170,173]). Unfortunately, the series of (fluoroalkyl)silanes investigated was incomplete [138], as is apparent from Table XVIII, which reproduces all results (except ^{1}H NMR data). The study was motivated by theoreti-

Table XVII. ^{1}H - NMR data on H_3SiCH_2Y compounds[a]

Y	$\delta(SiH_3)$	$\delta(CH_2)$	$^3J(^1H-^1H)$	$^1J(^{29}Si-^1H)$	$^1J(^{13}C-^1H)$
Cl	3.81	2.94	3.5	207.6	146.8
Cl[b]	4.03	2.96	3.7	206.3	141.6
Br	4.06	2.58	4.0	207.6	150.0
I	4.51	2.06	4.0	205.8	149.0

[a] Data converted from Ref. [169], measured in cyclohexane.
[b] Data from Ref. [170], $\delta(^{29}Si) = -56.48$, $\delta(^{13}C) = 20.7$.

cal considerations [174] according to which the "through-space" interaction between silicon and the functional group Y decreases in the order $NH_2 > OH > F$ and the "through-bonds" interaction decreases in the reverse order. As we have seen, studies of (amino- and hydroxyalkyl)-silanes have failed to reveal any such interaction; thus, it was hoped that the through-bond interaction could be reflected in the NMR spectra of (fluoroalkyl)silanes.

On brief inspection of Table XVIII it is obvious that most of the data on the α-derivative, (fluoromethyl)trimethylsilane, stand out as different from the remaining data. The difference is most striking for the ^{19}F chemical shifts; fluorine in the α-derivative is shielded some 55 ppm more than in the other (fluoroalkyl)silanes. (Similar shift values have been reported for the CH_2F fluorine in several other fluorinated derivatives of tetramethylsilane [175]). In a broader sense this fluorine chemical shift is not anomalous; it is about the same as that in methylfluoride, and it roughly fits the dependence on substituent electronegativity and net charge [138]. Moreover, anomalously high fluorine shielding would be difficult to associate with the α-effect. The small value of $^1J(^{19}F-^{13}C)$ coupling agrees with the assumed role of s-character of the bonding orbital and with Bent's rehybridization rules [176]. The ^{13}C chemical shifts fit the

Table XVIII. ^{29}Si, ^{19}F, and ^{13}C - NMR data on (fluoroalkyl)silanes[a]

Compound	δ(Si)	δ(F)	J(Si-F)	$^1J(F-C)$	$^2J(F-C)$	$^3J(F-C)$	$^2J(F-H)$
$(CH_3)_3SiCH_2F$[b]	-1.49	-270.37	21.0	161.0	-	2.2	47.0
$Cl_3Si(CH_2)_2F$	9.64	-210.51	28.2	171.3	22.1	-	47.0
$(CH_3)_3Si(CH_2)_3F$	2.16	-215.09	1.0	169.9	19.8	4.4	48.0
$Cl_3Si(CH_2)_3F$	13.34[d]	-219.85		169.9	21.3	5.1	46.5
$(CH_3)_3Si(CH_2)_5F$	1.45	-218.06	-				48.0
$n-C_7H_{15}F$	-	-218.19	-	164.8	19.5	4.9	47.5
$(CH_3)_3CCH_2F$	-	-222.23	-	173.5	17.69	4.4	48.0

[a] Data from Ref. [138], chemical shifts are in δ-scale. The shifts of ^{29}Si, ^{13}C, and ^1H nuclei are relative to TMS; the shifts of ^{19}F are relative to CFCl$_3$. Coupling constants are in Hz; the signs of the coupling constants have not been determined. The nuclei are identified by their symbols only; Si, F, C, and H stand for ^{29}Si, ^{19}F, ^{13}C, and ^1H nuclei, resp. Estimated errors are: 0.10 ppm for ^{29}Si and ^{13}C chemical shifts, 0.05 ppm for ^{19}F chemical shifts, 0.2 Hz for J(Si-F) and J(F-H) and 1.2 Hz for J(C-H) coupling constants. [b] Coupling constants of protons determined from ^1H NMR spectrum of the neat compound; $^2J(F-H)$ = 47.53 ± 0.07 Hz. $c^3J(C-H)$ = 5.9 Hz (^{13}C in CH_2F group; the ^{13}C of the $SiCH_2$ group is not seen in the spectrum).

Table XVIII continued

^{29}Si, ^{19}F, and ^{13}C - NMR data on (fluoroalkyl)silanes[a]

Compound	CH$_3$		Si-CH$_2$		C-CH$_2$-C		CH$_2$-F	
	δ(C)	J(C-H)	δ(C)	$^1J(C-H)$	δ(C)	$^1J(C-H)$	δ(C)	$^1J(C-H)$
(CH$_3$)$_3$SiCH$_2$F[b]	-4.25	119.44	80.03	137.1	–	–	80.03	137.1
Cl$_3$Si(CH$_2$)$_2$F	–	–	26.45	124.3	–	–	77.77	153.3[c]
(CH$_3$)$_3$Si(CH$_2$)$_3$F	-1.96	116.9	11.33	115.8	24.79	123.9	84.88	143.5
Cl$_3$Si(CH$_2$)$_3$F	–	–	19.85	120.2	23.39	e	83.10	150.6
(CH$_3$)$_3$Si(CH$_2$)$_5$F								
n-C$_7$H$_{15}$F			g		f		83.91	
(CH$_3$)$_3$CCH$_2$F	25.45						92.32	139.2[h]

[d]From 80% (v/v) solution in CCl$_4$. [e]Not determined from the complex spectrum. [f]Carbons at β and γ positions occur at δ = 30.93 and 25.59, resp., other resonances have not been assigned. [g]Tertiary carbon is found at δ = 32.37 but a line of 2-methyl-1-butene impurity also resonates in this region. If the given assignment were wrong, the shift would be δ = 31.40 and the coupling constant 31.6 Hz. [h]Determined from one satellite only, error ± 0.5 Hz.

Table XIX. ^{29}Si and ^{13}C NMR chemical shifts in $X_nMe_{(3-n)}Si(CH_2)_mY$ compounds[a]

X	Y	m	n	^{29}Si δ	^{29}Si Δδ	CH_3Si	$CH_3(CH_2O)$	CH_2O	$SiCH_2$	$β-CH_2$[b]	$γ-CH_2$[b]	$C=O$	$CH_3(CO)$
Cl	Cl	1	0	1.7	1.7	−4.0			29.8				
			1	22.9	−7.0	−0.3			30.1				
			2	21.7	−10.1	3.1			31.0				
			3	0.8	−11.4				30.8				
		2	0	−0.4	−2.0	−2.2			22.5	42.0			
			2	28.1	−5.9	5.5			26.2	38.8			
			3	8.3	−6.3				29.3	38.0			
		3	0	1.5	0.8	−1.6			14.6	28.0	47.6		
			1	31.3		2.0			17.0	27.3	47.5		
			2	32.7		6.0			19.9	26.9	47.2		
			3	12.4	−0.4				22.2	26.1	45.9		
OC_2H_5	Cl	1	0	1.7	1.7	−4.0			29.8				
			1	8.9	−4.6	−4.5	17.7	57.8	28.4				
			2	−17.2	−11.1	−7.2	17.6	58.0	26.0				
			3	−59.7	−15.2		17.3	58.1	22.9				

Table XIX continued

^{29}Si and ^{13}C NMR chemical shifts in $X_nMe_{(3-n)}Si(CH_2)_mY$ compounds[a]

X	Y	m	n	^{29}Si δ	Δδ	CH_3Si	$CH_3(CH_2O)$	CH_2O	$SiCH_2$	$\beta\text{-}CH_2$[b]	$\gamma\text{-}CH_2$[b]	C=O	$CH_3(CO)$
OC_2H_5	Cl	2	0	-0.4	-2.0	-2.2			22.5	42.0			
			2	-12.1	-4.8	-5.0	17.8	57.6	20.6	41.0			
			3	-52.3	-6.4		17.5	57.5	17.4	40.5			
		3	0	1.5	0.8	-1.6			14.6	28.0	47.6		
			1	14.8	0.0	-2.0	18.9	58.3	14.4	27.5	47.6		
			2	-7.1	0.5	-4.6	18.9	58.4	12.2	27.4	47.8		
			3	-47.0	0.0		18.7	58.7	8.7	27.4	47.5		
$OC(O)CH_3$	Cl	1	0	1.7	1.7	-4.0			29.8			170.2	21.3
			1	14.9	-6.9	-4.3			28.1			169.8	21.8
			2	-9.4	-14.0	-5.0			26.6			168.1	21.0
			3	-57.6	-13.9				23.7				

[a] Based on data of Refs. [47,142,177], maximum error ± 0.3 ppm.
[b] Position relative to the silicon atom.

dependence on the total net charge (Fig. 4) as satisfactorily as do similar carbon compounds, and the shifts also correlate linearly with $^1J(^{13}C-^1H)$ couplings (Fig. 9). The negative ^{29}Si SCS value in the α-derivative has already been explained (Section 3.3.3) on the basis of Del Re calculations as a consequence of the polarization of C-Si bonds similar to oxygen-containing carbon-functional compounds. More elaborate calculations [174] of charge distributions, which were performed only for model (fluoroalkyl)silanes and not for (fluoroalkyl)trimethylsilanes, do not support the above interpretation [138]. Typical data on (chloroalkyl)silanes are collected in Table XIX.

(Halomethyl)silanes undergo a rearrangement that can be described by the following equation

$$XCH_2SiR^1R^2_2 \longrightarrow R^1CH_2SiXR^2_2$$

where R^1 is the less electronegative of the substituents. Following this reaction (X=Cl, R^1=H) for 8 months by ^1H NMR, Bellama and Morrison [178] could prove that the rearrangement proceeds through the reaction sequence

$$2\ ClCH_2SiCl_nH_{3-n} \longrightarrow ClCH_2Si\underset{Cl}{\overset{H}{\diamondsuit}}CH_2SiCl_nH_{3-n} \longrightarrow$$

$$CH_3SiCl_nH_{3-n} + ClCH_2SiCl_{n+1}H_{2-n}$$

Other reactions of (haloalkyl)silanes were also followed by NMR, e.g., reaction with SbF_5 [179], with HSO_3F [180], with CF_3COONa [14]; and the relative reactivities of various C-H bonds in photochlorination were correlated with $^1J(^{13}C-H)$ coupling constants [181].

Ethene Derivatives. Interpretation of chemical and spectral properties of vinylsilanes also requires either $(p-d)_\pi$ bonding or σ-π interaction in addition to the inductive effect (see e.g., [182]). Similar interactions have also been invoked in allylsilanes, despite the fact that a methylene group separates the double bond from the silicon [183].

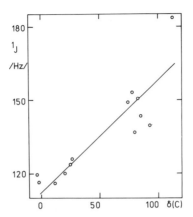

Fig. 9. Correlation of one-bond $^1J(^{13}C-^1H)$ coupling constant with carbon chemical shifts.

The ^1H chemical shifts of allylic protons (derived by a very approximative analysis of allylic proton spectra) and of methyl protons show a linear correlation with the sum of polar substituent constants $\sum \sigma^*$ of the substituents on the silicon. The inductive effect obviously plays a dominating role [184,185]. Deviations that occur for some methyl and methylene proton chemical shifts can be explained by: (a) magnetic anisotropy of phenyl rings and Si-halogen bonds, (b) by $(p-d)_\pi$ interaction between silicon and oxygen, and (c) by interaction of the allylic groups in di- and triallyl silanes. Deviations that occur in correlations of olefinic protons could be explained similarly and in some cases also by the formation of a cyclic intermediate:

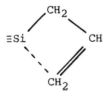

The chemical shift of the terminal olefinic proton in the cis position to the $-CH_2$-allylic group also shows a linear correlation with the wavenumber of the rocking vibration of the terminal olefinic CH_2 group [186]. It has been concluded that the extent of $(p-d)_\pi$ and $\sigma-\pi$ interaction depends on the nature and number of other silicon substi-

Table XX. ^{29}Si and ^{13}C – NMR chemical shifts in allylsilanes of the type $(CH_3)_{3-n}X_nSiCH_2CH=CH_2$ [a]

Substituent		Chemical Shifts $(CH_3)_{3-n}SiCH_2CH=CH_2$ group				Substituent X	
X	n	$\delta(^{29}Si)$	$\delta(^{13}CH_2)$	$\delta(^{13}CH)$	$\delta(=^{13}CH_2)$	$\delta(^{13}CH_3)$	$\delta(^{13}CH_3)$	Other
–	0	–0.4	24.4	133.8	112.1	–2.4	–	–
			25.4[b]	135.3[b]	113.5[b]	–1.5[b]		
			24.72[c]	135.08[c]	112.65[c]			
Cl	1	27.2	26.5	131.9	115.1	0.9	–	–
			28.3[b]	133.0[b]	116.3[b]	2.4[b]	–	–
			26.42[c]	131.98[c]	115.23[c]			
	2	26.8	27.7	128.5	116.7	3.3	–	–
			29.6[b]	130.6[b]	118.9[b]	5.2[b]	–	–
			28.59[c]	129.42[c]	117.54[c]			
	3	8.0	30.7	127.1	119.4	–	–	–
			3.6[b]	128.1[b]	120.6[b]	–	–	–

Table XX continued

^{29}Si and ^{13}C - NMR chemical shifts in allylsilanes of the type $(CH_3)_{3-n}X_nSiCH_2CH=CH_2$ [a]

Substituent X	n	δ(^{29}Si)	Chemical Shifts $(CH_3)_{3-n}SiCH_2CH=CH_2$ group δ($^{13}CH_2$)	δ(^{13}CH)	δ(=$^{13}CH_2$)	δ($^{13}CH_3$)	Substituent X δ($^{13}CH_3$)	other
			30.76[c]	127.24[c]	119.54[c]			
$(CH_3)_3SiO$	1	4.1	26.8	134.2	113.7	0.1	2.1	δ(^{29}Si)=7.3
	2	-26.5	26.1	134.0	114.1	-0.6	2.1	δ(^{29}Si)=6.9
	3	-70.3	22.6	133.7	114.1	—	1.8	δ(^{29}Si)=7.3
CH_3CH_2O	3	-51.6	18.1	132.8	113.7	—	58.0	δ($^{13}CH_2$)=17.8
			18.08[c]	132.41[c]	114.28[c]			
CH_2CHCH_2	1	0.2	23.1	134.6	113.6	-3.8		
			22.71[c]	134.74[c]	113.00[c]			
$(CH_3)_3SiCH_2$ [d]	—	1.0	28.1[c]	140.4	112.2	-1.8		

[a] Data from ref. [182]. [b] Data from ref. [188]. [c] Data from ref. [187]. [d] Data for $(CH_3)_3SiCH_2CH_2CH=CH_2$. [e] The values for allylic CH_2 group, δ($^{13}CH_2Si$) = 15.8.

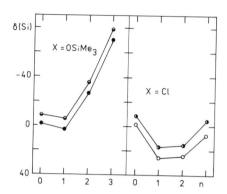

Fig. 10. ^{29}Si chemical shift dependence on the number (n) of X substituents in vinylsilanes $C_2H_3SiX_n(CH_3)_{3-n}$ (open and full points) and in allylsilanes $C_3H_5SiX_n(CH_3)_{3-n}$ (halves points).

tuents. Contrary to these conclusions, ^{1}H NMR spectra of a small series of allylsilanes (but including allyl derivatives of other group IV elements) did not lend any support (when both chemical shifts and coupling constants were considered) to the existence of $(p-d)_\pi$ overlap in the ground state [183].

Since ^{13}C chemical shifts of olefinic carbons show a clear relationship to the electron density, it was hoped that a more definite solution would result from measurements of ^{13}C NMR spectra. The measurements were carried out simultaneously by Rakita and Worsham [187] and in this laboratory [182] where the ^{29}Si NMR spectra were also measured. The results are summarized in Table XX and the trends are illustrated in Figs. 10 and 11. The chemical shifts of olefinic carbons are similar to those in branched alkenes. The shifts of the γ-carbon of CH_2 and the β-carbon of CH groups are linearly related [182,187]; they fit reasonably well the general regression described for allylic compounds [189], but they do not correlate with the shifts of the connecting CH_2 α-carbon. The latter shift correlates with the corresponding methyl shifts in $RSiX_3$ compounds [187]. All of these carbon chemical shifts can be predicted to within ±0.16 ppm by a pairwise additivity scheme (the increments are found in Ref. [187] for the following substituents on the silicon atom: CH_3, C_3H_5,

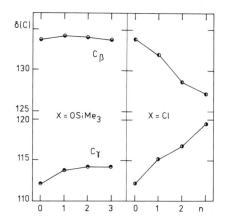

Fig. 11. ^{13}C chemical shift dependence on the number ($\underline{n}$) of X substituents in allylsilanes $C_3H_5SiX_n(CH_3)_{3-n}$ (carbon position relative to the silicon atom).

C_6H_5, Cl, OR, and H). The internal shift between the two olefinic carbons is an approximate measure of the polarity of the double bond. In allyltrimethylsilane the internal shift indicates polarity in the sense

$$\begin{array}{cc} \delta- & \delta+ \\ CH_2=CH- & \end{array}$$

Increasing substitution of methyl groups by more electronegative groups decreases the polarity of the terminal olefinic group. Surprisingly, the shielding of the olefinic carbons in allylsilanes is about equally sensitive to that in vinylsilanes with respect to substitution on the silicon. The CH_2 olefinic carbons, in accord with the less pronounced delocalization in allylsilanes, are shielded 10-18 ppm more than in vinylsilanes.

The silicon-29 in allylsilanes is shielded less than in vinylsilanes (where the increased shielding is supposedly due to $(p-d)_\pi$ interaction) but more than in ethyl- or propylsilanes. It is possible that the higher shielding is related to π-d or σ-π interaction. The dependence of the ^{29}Si SCS on substitution on the silicon atom shows once again the algebraic increase of negative SCS values with increasing $\underline{n}$, which is caused by a polarization of Si-X

bonds under the influence of the vinyl group that is more
electronegative than the methyl group in the parent
compound.

The different substituent effects of the $Si(CH_3)_3$ and
$CH_2Si(CH_3)_3$ groups also have their counterparts when these
groups are attached to a benzene ring instead of to a
vinylic system.

Benzene Derivatives. Of all possible carbon-
functional compounds with $Y = C_6H_5$, interest has been con-
centrated almost exclusively on compounds with m = 1,
i.e., on benzylsilanes. First, NMR studies can help to
elucidate different substituent effects of trimethylsilyl
and trimethylsilylmethyl groups on a benzene ring; second,
NMR studies promise to allow differentiation among inductive effects and π-d and π-σ* interactions through the
different conformational dependence of the latter two
interactions; and, finally, the magnetic non-equivalence
of benzylic protons provides an interesting NMR problem of
its own.

The dominating influence of the α-substituents on the
^{29}Si chemical shifts is reflected in a good linear correlation between the chemical shifts in benzylsilanes ($Y = C_6H_5$;
m = 1) and in the parent (Y = H; m = 1) compounds [141] and
also in an exceedingly good correlation with the shifts in
the corresponding phenylsilanes (eleven data points, r =
0.999) [141].

$$\delta[^{29}Si(benzylsilane)] = 7.48 + 1.01 \, \delta[^{29}Si(phenylsilane)] \quad (18)$$

The unit slope shows that hybridization of the carbon atom
attached to the silicon does not affect the trend due to
the α X substituents. The intercept, 7.48 ppm, represents
the shift of the resonance in benzylsilanes to lower field
compared to the resonance of the corresponding phenylsilanes. According to the quality of correlation, the value
of 7.48 ppm is a constant that does not depend on the nature
and number of substituents X, and it also does not depend
on which branch (A or B in Fig. 2) on which the corresponding point is located as determined by the X substituents.
An interpretation of this observation has not yet been
offered (it has only been shown that the contribution from
magnetic anisotropy can be neglected [141]). Without
invoking the σ_E term, the ERW theory can hardly explain

how the replacement of $CH_2C_6H_5$ by a C_6H_5 group can increase the electron density in one branch and decrease it in the other. It might at first appear an easy task to explain this observation in terms of the qualitative model: i.e., $(\pi-d)_\pi$ bonding between the aromatic ring and the silicon atom in phenylsilanes can be made responsible for this additional shielding in phenylsilanes compared to the shielding in benzylsilanes. However, numerous other observations have shown that acceptance of the idea of $(\pi-d)_\pi$ bonding between the aromatic ring and the silicon atom leads necessarily to the conclusion that the extent of this bonding depends on the nature and number of the other silicon substituents.

The $^1J(^{29}Si-^1H)$ coupling constants in substituted benzyldimethylsilanes show good linear correlation with Hammett σ constants of meta- and para- substituents [190], a situation analogous to the similar correlations for phenylsilanes (see section 3.4.2) in which the sensitivity to substituent effects is about the same as in phenyltetramethyldisilanes. This coupling is appreciably more sensitive to substituent effects than the $^1J(^{13}C-^1H)$ coupling in the corresponding β C-H bond (relative to the benzene ring) in the above mentioned series of compounds [190].

Chemical shifts of protons and carbons of the methyl groups attached to the silicon atom show the usual dependence on the nature and number of the substitutents [141]. Proton chemical shifts in trimethylsilylmethyl compounds are of some interest. The following shifts were reported [141,191]:

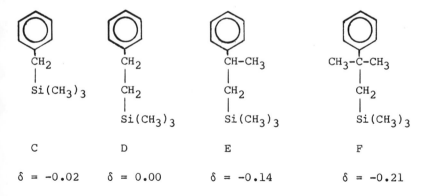

C	D	E	F
δ = -0.02	δ = 0.00	δ = -0.14	δ = -0.21

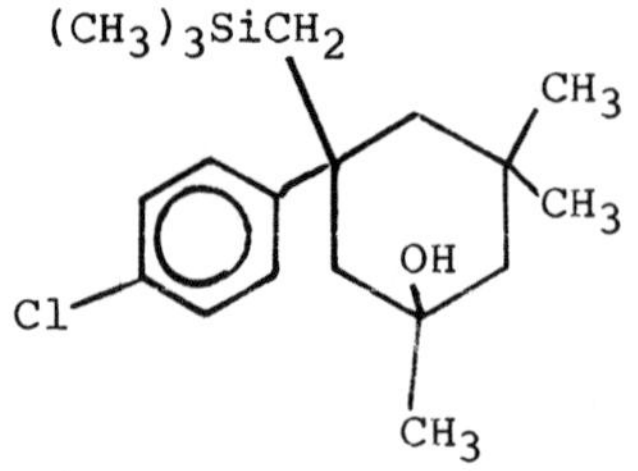

(CH₃)₃SiCH₂ structure with cyclohexane ring bearing CH₃, CH₃, OH, CH₃ substituents and a 4-chlorophenyl group:

$$\text{G}$$

$$\delta = -0.36$$

The unusual upfield shifts were explained [191] by the predominance (on a time-averaged basis) of conformations, which place the methyl groups in the face of the aromatic ring.

The diastereotopic methylene protons adjacent to the silicon in compounds E and G have large chemical shift differences (0.47 and 0.10 ppm, respectively, at 31°) [191].

The interesting stereochemical problems of magnetic nonequivalence are not specific for carbon-functional compounds with $Y = C_6H_5$, but in these compounds the chemical shift nonequivalence is large, and it sometimes leads to directly observable intrinsic chemical nonequivalence. Thus, Redl and Peddle [192] measured temperature dependence of the shift difference in the benzylic methylene protons in

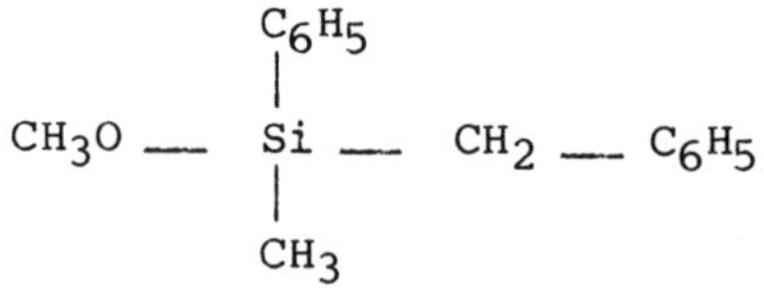

$$CH_3O - \underset{\underset{CH_3}{|}}{\overset{\overset{C_6H_5}{|}}{Si}} - CH_2 - C_6H_5$$

H

Above 160° the high-temperature limit was reached, conformer populations were approximately equal, and the chemical shift difference reached a constant value of 0.071 ± 0.002 ppm (intrinsic diastereotopic nonequivalence).

The chemical shift nonequivalence of protons and carbons of the two methyl groups attached to the silicon atom in compounds of the type

```
        H              CH₃
        |              |
CH₃  -  C  -  CH₂  -  Si  -  R
        |              |
        C₆H₅           CH₃
```

I

was also studied [193,194]. For example, in the case
R = C$_6$H$_5$ the difference in carbon chemical shifts is 0.5 ppm
[194] and in proton shifts only 0.049 ppm [193].

In general, the methylene proton and carbon chemical
shifts in benzyl derivatives show "normal" linear dependences on substituent electronegativity (i.e., a decrease in
shielding with increasing substituent electronegativity)
[195], and the two shifts correlate linearly with each
other [141]. In benzylsilanes, C$_6$H$_5$CH$_2$Si(CH$_3$)$_{3-n}$X$_n$ (X = F,
Cl, OC$_2$H$_5$) [141], only the shifts in the chloro derivatives
follow these general trends (corresponding to an increasing
group electronegativity with increasing $\underline{n}$). Methylene carbon shieldings in compounds with X = F or OC$_2$H$_5$ increase with
increasing $\underline{n}$; the proton shifts follow the general trend in
fluoro derivatives but remain essentially constant in
ethoxy derivatives.

Aromatic carbon ^{13}C SCS values for several
(CH$_2$)$_m$SiX^{1}X^{2}X^3 groups are given in Table XXI. The order of
sensitivity to substituent effects in benzyl derivatives is
C-1 > C-4 > C-2 > C-3, and the range of chemical shifts is
smaller than in phenylsilanes. The C-4 carbon chemical
shifts in benzylsilanes show good linear correlation with
the corresponding shifts in phenylsilanes. The average
difference between the two shifts (6.0 ppm) corresponds to
the total charge density; it is 0.02 units more negative on
the C-4 carbon in benzylsilanes than in phenylsilanes, and
the σ^+ constant of the CH$_2$SiX^{1}X^{2}X^3 group is 0.5 units more
negative than that of the SiX^{1}X^{2}X^3 groups [141].

All of these observations are in agreement with silicon acting in phenylsilanes as an electron acceptor and the
CH$_2$SiX^{1}X^{2}X^3 group as a donor.

It was suggested by Eaborn [197] in 1956 that the
electron releasing ability of the CH$_2$Si(CH$_3$)$_3$ group is due

Table XXI. Substituent effects on aryl carbon-13 chemical shifts in I-X-substituted benzenes[a]

Substituent X	Aryl carbon position			
	C - 1	ortho	meta	para
$Si(CH_3)_3$	+11.3	+4.7	-0.8	+0.2
$Si(CH_3)_2Cl$	+8.1	+5.1	+0.2	+2.2
$Si(CH_3)Cl_2$	+4.1	+4.3	-0.4	+2.9
$SiCl_3$	+3.3	+5.0	+0.5	+4.7
$Si(CH_3)_2F$	+7.9	+4.8	-0.2	+2.0
$Si(CH_3)F_2$	+1.6	+4.9	-0.1	+3.4
$CH_2Si(CH_3)_3$	+11.0	-0.9	-0.9	-5.0
$CH_2Si(OC_2H_5)_3$	+8.8	-0.1	-0.9	-4.5
CH_2SiCl_3	+3.0	+0.4	+0.4	-2.1
SiF_3	-8.0	+5.8	0.0	+4.6
SiH_3	-0.6	+7.4	-0.4	+1.3
$Si(CH_3)_2(OC_2H_5)$	+9.7	+4.9	-0.7	+0.9
$Si(CH_3)(OC_2H_5)_2$	+5.7	+4.8	-1.4	+0.7
$Si(OC_2H_5)_3$	+3.2	+6.1	-0.9	+1.4
$Si(C_6H_5)_3$	+6.4	+8.5	0.0	+1.8
$SiH_2(C_6H_5)$	+3.5	+7.9	+0.3	+2.1
$Si(CH_3)(C_6H_5)_2$	+8.3	+7.5	-0.1	+1.5

Table XXI continued

Substituent effects on aryl carbon-13 chemical shifts in
I-X-substituted benzenes[a]

Substituent X	Aryl carbon position			
	C - 1	ortho	meta	para
$CH_2Si(CH_3)_2Cl$	+7.8	-0.5	-0.5	-4.0
$CH_2SiCH_3Cl_2$	+5.5	+0.1	+0.1	-2.8
$CH_2Si(CH_3)_2(OC_2H_5)$	+10.3	-0.6	-0.6	-4.7
$CH_2Si(CH_3)(OC_2H_5)_2$	+9.5	-0.4	-0.4	-4.5
$CH_2Si(CH_3)_2F$	+8.8	-0.2	-0.2	-4.0
$CH_2SiCH_3F_2$	+5.5	-0.3	-0.3	-3.5
CH_2SiF_3	+2.9	+0.3	+0.3	-2.3
$CH_2CH_2Si(CH_3)_3$	+16.9	0.0	0.0	-2.4
$CH_2CH_2CH_2Si(CH_3)_3$	+14.3	+0.5	+0.5	-2.2
$(CH_2)_4Si(CH_3)_3$	+14.8	+0.7	+0.7	-1.9

[a] Data taken from Ref. [196]. In ppm relative to external benzene, positive values indicate deshielding. Values derived for neat liquids, reliable to ± 0.5 ppm.

to hyperconjugative release from the CH_2-Si bond in addition to the inductive release of electrons to carbon by the $Si(CH_3)_3$ group. The relative importance of the two mechanisms was probed by ^{19}F and ^{13}C NMR spectroscopy (among other techniques). According to the trends in ^{19}F chemical shifts in $p\text{-}F\text{-}C_6H_4\text{-}CH_{3-x}(Si(CH_3)_3)_x$, studied by Bassindale et al. [198], the electron release cannot be attributed entirely or even primarily to inductive release. The attempt to assess the relative importance of these mechanisms quantitatively from a DSP analysis of ^{19}F chemical shifts (according to Eq. 29) was not conclusive [199]. (The hyperconjugative nature of the effect of CH_2SiR_3 groups in other aromatic systems was also confirmed by ^{19}F and ^{13}C NMR [200,201]). The conformational dependence of ^{19}F chemical shifts in bicyclic silacyclic compounds was taken [202] as indicating the predominance of the π-σ mechanism. However, in the bicyclic compounds the aromatic ring is somewhat distorted, and as there are also other objections to using a fluorine tag for such investigations (see next section), the results of ^{13}C NMR studies seem to be more reliable. It was found [203] that the electron-releasing effect of the $CH_2Si(CH_3)_3$ group is suppressed when the optimum alignment of the C-Si bonds and the system is prevented. Other ^{13}C NMR studies confirmed (through the conformational dependence of the chemical shifts) the importance of the hyperconjugative mechanism in benzylic compounds [204-206].

The same π-σ interaction explains the high ability of the $CH_2Si(CH_3)_3$ group to delocalize a positive charge, as illustrated by the ^1H NMR spectra of protonated 1,3-dimethyl-5-trimethylsilylmethylbenzene and 1-methyl-3-trimethylsilylmethylbenzene, in which the proton is associated to the benzene ring in the para position (relative to the trimethylsilylmethyl group) [207].

3.4 AROMATIC CARBON-FUNCTIONAL COMPOUNDS

As in other branches of chemistry, compounds containing a benzene ring are one of the most frequently studied types of carbon-functional compounds. In the case of NMR studies several factors have historically contributed to this popularity: (a) the high analytical potential of NMR for structural determination of these important compounds, (b) the large number of NMR parameters that can be measured and for which methods of interpretation have been established in other studies, and (c) a theoretical interest in these compounds.

It is rather unfortunate that nearly all research concerning this class of compounds can be divided into two distinct groups: (1) the studies of the effects of various silyl groups on functional group Y, and (2) the effects of the Y substituents on $SiX^1X^2X^3$ groups. However, a third group unites both aspects and also follows the effects on the transmitting C_6H_4 chain. Only a few papers are relevant to this third category.

3.4.1 NMR Spectroscopy of Functional Groups

Measurements of NMR spectra of the Y functional groups are usually aimed either at the determination of various substituent constants of $SiX^1X^2X^3$ silyl groups or at the evaluation of the relative efficiency of silicon to transmit the electronic effects of the X substituents.

Since substituent constants are parameters designed to characterize the effects of substituents on chemical reactions (on the equilibrium or kinetic rate constants), there are many reasons to expect linear correlations between pertinent substituent constants and those NMR parameters that measure chemical equilibrium or rate. This is, e.g., the case with phenolic (Y = OH) or carboxylic (Y = COOH) proton chemical shifts in dry solvents. The phenolic proton chemical shift (in dimethylsulfoxide) yielded the value of the σ-constant of the trimethylsilyl group [208] (see Table XXII). The carboxylic proton shifts in dry pyridine were measured by two research groups [212,219, 220]. Correlation of this chemical shift with the Hammett σ constant [221] allowed estimation of the substituent constant for several silyl groups. Although the experimental results are extremely sensitive to traces of water (δ = 5.5), which reduce the chemical shift of the carboxylic proton (δ = 14) by proton exchange (and thus increase the values of the substituent constants), the precision achieved permits a limited comparison of the values obtained for very similar substituents (m- and p-trimethylsiloxymethylsilyl, i.e., $((CH_3)_3SiO)_{3-n}(CH_3)_nSi$ groups).

Table XXII. Examples of substituent constants determined from NMR data on carbon-functional compounds

Substituents	Constants		Functional group Y	Reference
$(CH_3)_3Si$	$F = -0.26_6$	$R = +0.09_5$	$CH=CH_2$	[209]
	$\sigma_I = -0.09$	$\sigma_R^o = +0.07$	$CH=CH_2$	[206]
	$\sigma_I = -0.10$	$\sigma_R^o = 0.06$	F	[210]
	$\sigma_I = 0.02$	$\sigma_R^o = 0.05$	F	[214]
	$\sigma_I = 0.01$	$\sigma_R^o = 0.04$	F	[215]
	$\sigma_I = 0.00$		F	[216]
	$\sigma_p = -0.05$		F	[211]
	$\sigma_p = -0.06$		COOH	[212]
	$\sigma_p = 0.06$		CH_3	[213]
	$\sigma_p = -0.21, -0.11, -0.15$		OCH_3	[217]
	$\sigma_p = -0.37$		CH_3	[217]
	$\sigma_p = -0.02$		COOH	[212]
	$\sigma^- = 0.06$		OH	[208]
$(CH_3)_3SiCH_2$	$\sigma_I = -0.10$	$\sigma_R^o = -0.15$	$CH=CH_2$	[206]
	$\sigma_R^o = -0.20$		F	[218]
	$\sigma_I = -0.076$	$\sigma_R^b = -0.200$	F	[199]

Hammett type correlations for differential solvent shifts fall in between the correlations with and without an obvious relationship to chemical reactions. Several such correlations were employed for estimation of substituent constants of the $Si(CH_3)_3$ group [217]. For a selection of derived values see Table XXII.

Linear correlations between NMR parameters on the one hand and single (SSP) or "dual" (DSP) substituent parameters on the other hand have been reported for a number of other situations (see reviews [222,223]) in which the NMR parameter is not related to any chemical reaction. Justification and theoretical interpretation of such correlations are the subject of continuing discussion (e.g., [222-224]). If such a correlation is satisfactory in a statistical sense (i.e., if it can yield sufficiently precise estimates of substituent constants), then it is likely to be used for the determination of substituent constants and will eventually also be used on organosilicon groups.

The extensive literature on substituent effects in the NMR spectra of disubstituted benzenes contains a large amount of scattered information on various organosilicon compounds. For example, substituent effects on ^1H or ^{13}C NMR spectra of 4-substituted styrenes [209], α-methylstyrenes [225], phenylacetals [226], acetanilides [227], and dimethylaminochlorophenylboranes [228] were studied on series of compounds that included trimethylsilyl derivatives. Specific information can be obtained from compound registers [171]; of concern here will be only the more systematic studies.

Before ^{13}C NMR spectrometers became generally available, the most popular LFE correlations in NMR were those developed by Taft and his group [218,229] for ^{19}F chemical shifts in substituted fluorobenzenes. The impetus to their work was provided by the early finding of Gutowsky et al. [230] of the extreme sensitivity of fluorine shielding to substituent effects in fluorobenzenes, as expressed by the large regression coefficients in a rough correlation with Hammett constants. (This crude correlation was recently also utilized by Varlamov et al. [231] for estimation of the Hammett constants of a few $RC(O)OSi(CH_3)_2$ groups). Detailed analysis of the contributions to fluorine shielding in para and meta substituted fluorobenzenes led [218,229] to a system of two linear equations for substituent chemical shifts

Table XXIII. ^{19}F SCS values determined in CCl_4 solutions by different workers

SiX_3 group	SCS_{para}	SCS_{meta}	Reference
$Si(CH_3)_3$	0.51	0.83	[238]
	0.50		[218]
	0.50	-0.85	[234]
SiF_3	9.57	2.35	[236]
	9.010	1.865	[237]
$SiCl_3$	7.90	2.10	[234]
	7.594	1.716	[236]
$SiBr_3$	7.55	2.15	[234]
	7.260	1.775	[236]

$$SCS(X)_{meta} = 7.1\,\sigma_I - 0.60 \qquad (29a)$$

$$SCS(X)_{para} = 29.5\,\sigma_R^o + SCS(X)_{meta} \qquad (29b)$$

Here $SCS(X)_{meta}$ and $SCS(X)_{para}$ represent the substituent chemical shifts of fluorine in the meta and para positions, respectively. The SCS values were measured in dilute carbon tetrachloride solutions relative to internal fluorobenzene. (In other papers from Taft's laboratory [232, 233] slightly different values for the numerical coefficients in Eq. (29) were reported.) Since the ^{19}F SCS values are solvent dependent, so also are the values of the numerical coefficients in Eq. (29). Significantly different SCS values were reported for measurements taken in the same solvent but in different laboratories (Table XXIII).

The fundamental assumptions on which the Taft treatment is based have been subjected to criticism by several groups of authors (see reviews [222,223,239]). It has been shown, e.g., that geometrical and conformational factors also affect the SCS values in fluorobenzenes [214]. Adcock et al. [214] have proposed using for similar correlations the SCS values measured in 2-substituted 6β- and 7β-fluoronaphthalenes

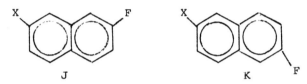

that do not appear to be complicated by substituent induced distortions.

Despite this criticism, Eq. (29) has often been used for derivation of substituent constants from measured SCS values. Many of the substituent constants that have been thus determined have been incorporated into critical collections of substituent constants (e.g. [240]), and yet Eq. (29) is designed for prediction of SCS values and not for evaluation of substituent constants. For that purpose the data of Taft et al. [218,229] should be fit by inverse equations that would minimize the deviations in substituent constants. Usually, authors simply follow the example set by Taft's group [218] and solve Eq. (29) for the unknown substituent constants; several such applications have included various silyl groups.

Thus, the work of Taft et al. [218] has already included some fluorobenzenes substituted by $Si(CH_3)_3$ and $CH_2Si(CH_3)_3$ groups. According to the ^{19}F shift in para-trimethylsilylfluorobenzene, the former substituent is of the +R type. The fact that there was no measurable solvent dependence observed for this compound was explained by the formal charge on the Si atom being buried within the molecular cavity.

$$\text{F—} \bigcirc \text{—Si(CH}_3)_3 \quad \longleftrightarrow \quad \overset{\delta+}{\text{F}} = \bigcirc = \overset{\delta-}{\text{Si(CH}_3)_3}$$

L   M

Similarly, the measurements of meta and para substituted fluorophenyldiphenylsilanes [241] yielded σ_I and σ_R^o constants of -Si(C$_6$H$_5$)$_2$X groups (X = NH$_2$, H, OH, OC$_2$H$_5$, Cl, F, N$_3$, Br). The derived values of the σ_R^o constants show that the -Si(C$_6$H$_5$)$_2$X groups are in all cases electron withdrawing because of resonance. It could not be decided which of the possible resonance forms (N, O, and P) was more important.

$$\text{F}^+ = \bigcirc = \text{Si}^- - \text{X} \quad\quad \text{F} - \bigcirc - \text{Si}^- = \text{X}^+ \quad\quad \text{F}^+ = \bigcirc = \text{Si}^- = \text{X}^+$$

N   O   P

Later, ^{31}P chemical shifts in compounds with X = NP(C$_6$H$_5$)$_3$ indicated that form O is important [242]. A comparison of shift ranges in meta derivatives suggests that silicon is better able to transmit substituent inductive effects than is carbon in analogous compounds [241]. In para isomers the situation is more complex. Interposition of a (C$_6$H$_5$)$_2$C group between the fluorophenyl tag and the triarylphospinimine group produces a deshielding of the para ^{19}F nucleus that can be attributed to a loss of resonance [242]. Interposition of a (C$_6$H$_5$)$_2$Si group into the same position produces a large deshielding effect, which can be explained by structure N and a utilization of vacant silicon acceptor orbitals [242].

These problems of benzene (or naphthalane) ring interaction with silicon (or with another group IVB atom) have been investigated by several other research groups using NMR of ^{19}F and of other nuclei; the most notable being the results of Maire and coworkers [238,243] and of Adcock and Kitching with their collaborators [214,215, 244-246]. The Si(CH$_3$)$_3$ group was found to be engaged in $(\pi-d)_\pi$ conjugative electron-withdrawal in p-fluorophenyl- [244], 4-fluoro-4'-biphenyl-[244], and 4α-, 6β-, and 7β-fluoronaphthyl- [214,245,246] trimethylsilanes.

Replacement of the CH_3 group by more electronegative groups like phenyl or vinyl increases the effective electronegativity of the entire group [244]. Replacement by hydrogen atoms also increases the electron-withdrawing power of the whole substituent, the origin of the change being in the polar term (σ_I) [215]. Substituents such as CH_2SiX_3 are effective ortho-para donors [244], as confirmed by other NMR results [196,215,247]. For a discussion of these results see section 2.3.4.

Other examples of substituent constant determination from ^{19}F chemical shifts include constants for SiX_3(X = H, F, Cl, and Br), as listed in Table XXIV. In view of the discrepancies illustrated by the data in Table XXIII, the agreement between the constants in Table XXIV is good.

Lipowitz [235] has exploited the high sensitivity of fluorine chemical shifts to substituent effects in a different way. Instead of evaluating substituent constants for about thirty $SiX^1X^2X^3$ groups for which he measured ^{19}F SCS_{para} and SCS_{meta} values (Table XXV), he correlated these quantities with the sum of the inductive constants of the X substituents, $\sum_i \sigma_I(i)$. Deviations from otherwise very good linear correlations occur for the first row

Table XXIV. Substituent constants for SiX_3 groups determined from ^{19}F SCS

SiX_3 groups	σ_I		σ_R^o	
SiH_3	0.09[a]		0.09[a]	
SiF_3	0.42[b]	0.41[c]	0.24[b]	0.24[c]
$SiCl_3$	0.39[b]	0.38[a]	0.17[b]	0.20[a]
$SiBr_3$	0.39[b]	0.39[a]	0.18[b]	0.18[a]

[a] Calculated according to Eq. (29) from data of Ref. [234], measured in CCl_4.
[b] Taken from Refs. [237,248], measured in n-heptane solutions.
[c] Calculated according to Eqs. (29) from data of Ref. [236], measured in CCl_4.

Table XXV. Some ^{19}F - SCS values for $F-C_6H_4-SiX^1X^2X^3$ compounds[a]

Substituents	X^1	X^2	X^3	SCS_{para}	SCS_{meta}
	F	F	F	9.32	-
	Cl	Cl	Cl	7.79	2.10
	CH_3	F	F	6.08	
	CH_3	Br	Br	5.56	
	CH_3	Cl	Br	5.53	1.18
	CH_3	Cl	Cl	5.52	1.11
	CH_3	CH_3	F	3.09	-0.06
	CH_3	CH_3	Br	3.23	
	CH_3	CH_3	CH_2Cl	2.07	-0.16
	CH_3	CH_3	$CHCl_2$	3.22	
	OCH_3	OCH_3	OCH_3	3.12	-0.59
	CH_3	CH_3	Cl	3.08	0.15
	CH_3	OCH_3	OCH_3	2.43	
	CH_3	CH_3	OCH_3	1.60	
	CH_3	CH_3	H	1.16	-0.72
	CH_3	CH_3	$CH_2C_6H_5$	0.98	
	CH_3	CH_3	C_6H_5	0.96	
	CH_3	CH_3	C_2H_5	0.43	
	CH_3	CH_3	C_2H_5	0.73	-0.82

[a] Data converted from Ref. [235] measured in 5% cyclohexane solutions relative to internal fluorobenzene standard.

donor atom substituents (N, O, F) attached to silicon. In such cases, the contributions from structure Q were throught to be significant.

Q R

The overall fit improved after SSP treatments were replaced by DSP analyses. From a comparison of para and meta SCS values Lipowitz [235] concluded that although mesomeric structure R contributes significantly to the SCS_{para} value, its contribution to the ground state is small.

The high sensitivity of fluorine to substituent effects also permitted statistically significant differentiation between the abilities of $Si-C_{arom}$ and $Sn-C_{arom}$ bonds to transmit substituent effects [249]. The ^{19}F chemical shifts in $p-F-C_6H_4-MAr_3$ (M = Si and Sn), in which Ar represents various meta or para substituted phenyl groups, correlate very well with the Taft inductive constants of the Ar groups [249]. On the basis of these correlations, as well as on the basis of DSP analysis, the authors [249] concluded that the electronic effects are transmitted through $Si-C_{arom}$ and $Sn-C_{arom}$ bonds mainly by an inductive mechanism, and on a 95% confidence level the transmission is more effective when M = Si than when M = Sn.

^{14}N NMR was also utilized to help to elucidate the bonding problems in phenylsilanes. The chemical shifts of all three isomeric trimethylsilylnitrobenzenes (ortho = +7, meta = +9, and para = +12) ppm suggest that there is some conjugation between silicon 3d orbitals and π-orbitals of the benzene ring [243].

Correlations that involve proton chemical shifts or coupling constants have also been applied to organosilicon compounds. Correlations valid for proton chemical shifts usually have smaller slopes than analogous correlations for other nuclei and also are substantially affected by the choice of solvent and concentration. In addition to the above mentioned correlations for benzoic and phenolic protons, the example of single substituent parameter correlations for methyl proton chemical shifts and ^{13}C-^{1}H coupling constants in sustituted toluenes [190,250] can also be noted. A judicious choice of solvent [213] permitted estimation of σ_p (±0.05) constants of $Si(CH_3)_{3-n}X_n$ groups (X = H, Cl, F, OCH_3). Methyl proton chemical shifts and ^{13}C-^{1}H coupling constants were also measured in substitued t-butylbenzenes, anisoles, and dimethylanilines [251]. The chemical shifts and coupling constants in anisoles and dimethylanilines, which were supported by the chemical shifts in toluenes, indicated that the order of effective substituent electronegativity of the para $M(CH_3)_3$ groups is Si > Ge > C [251]. The order was ascribed to a significant contribution of the following structure [251]:

$$CH_3O^+ = \langle \text{benzene} \rangle = Si^-(CH_3)_3$$

S

In meta derivatives, where such direct resonance interaction is not possible, the effective electronegativities of these groups are nearly equal [251].

The methods discussed above, which evaluate electronic interactions between a silicon atom and a benzene ring from the NMR effects observed on the Y functional group, suffer from an inadequacy noted by Katritzky et al. [252,253] for the case of Y = F; the Y substituent can interfere with the interactions, and thus the conclusion is not independent of the method employed. Of the methods that do not require the presence of any Y substituent, ^{13}C, ^{1}H, and ^{29}Si NMR spectroscopy of Si-substituted phenylsilanes should be noted, but such compounds obviously are not carbon-functional.

3.4.2 NMR Spectroscopy of $SiX^1X^2X^3$ Silyl Groups

The discussion of ^{29}Si chemical shifts and coupling constants will be followed by consideration of other nuclei as they occur at increasing distances from the benzene ring.

Though several authors had previously published the ^{29}Si chemical shifts of a few phenylsilanes [17,48,51,59, 196,254-256] and compared them with the data on analogous compounds of other elements, the definitive description of substituent effects on these shifts is due to Ernst et al. [52,257]. They extended the study of Maciel's group [51] of Hammett type dependence of the silicon shielding in $p-Y-C_6H_4-Si(CH_3)_3$ compounds to other silyl groups. Their results are summarized in Fig. 12. The sign of the ^{29}Si SCS value (or the slope ρ in the Hammett dependence) depends on the nature of the X substituents directly attached to the silicon atom. For substituents X = H and CH_3, the electron-withdrawing Y substituents cause deshielding of the silicon, while the same Y substituent brings about increased shielding if X = F, Cl, or OC_2H_5. Interpretation of these findings by the back-bonding model

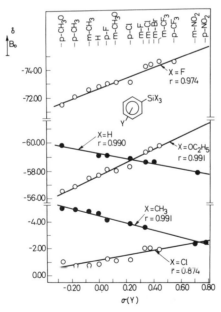

Fig. 12. ^{29}Si chemical shift dependence on Hammett σ constants of the Y substituents in substituted phenylsilanes (Adapted from [52]).

described in Section 3.2.3 is understandable if the following mesomeric structures are used to describe the ground state of the compounds considered:

$$\underset{T}{\underset{Y}{\bigcirc}-\overset{X}{\underset{X}{\overset{|}{Si}}-X}} \quad \longrightarrow \quad \underset{U}{\underset{Y}{\bigcirc}-\overset{X}{\underset{X}{\overset{|}{\bar{Si}}=X^+}}} \quad \longrightarrow \quad \underset{V}{\underset{Y}{\bigcirc}=\overset{X}{\underset{X}{\overset{|}{\bar{Si}}-X^+}}}$$

In compounds in which structure U can be significant (X = F, Cl, OC_2H_5), electron withdrawing Y substituents enhance the X→Si back-bonding. This back-bonding leads to an increase in the shielding, which overcompensates for the deshielding effect produced by polarization of the C_{arom}-Si bond in structure T. If structure U is not significant (X = H, CH_3), then only the latter effect operates. The authors [52] did not accept this explanation, since CNDO/2 calculations of electron density do not support it. (It has previously been mentioned, however, that these calculations are often unreliable[*] [77,78]). Instead, the authors adapted the Letcher-Van Wazer theory [79] of ^{31}P shielding to apply to silicon shielding. At present their theory can be considered as a special case of the ERW theory (for a detailed comparison with the corresponding stage of the ERW theory see [61]). As was demonstrated by Radeglia and Engelhardt [76], the ERW theory without inclusion of silicon 3d orbitals and the σ_E term can account very well for the discussed trends. The X substituents with high electronegativities (X = F, Cl, OC_2H_5) place the point corresponding to the given series on the B branch, while substituents such as a methyl group or a hydrogen atom place it on the A branch as shown in Fig. 2.

---

[*] On the other hand, according to this model, larger slopes could be predicted and were experimentally confirmed [81] for analogous dependences in $Y-C_6H_4-OSi(CH_3)_3$ compounds. In order to explain this observation, the ERW theory had to incorporate the σ_E term [73].

The one-bond or direct ^{29}Si-^{1}H coupling constants (which are negative) in meta and para substituted phenylsilanes, phenylmethylsilanes, and phenyldimethylsilanes exhibit excellent linear correlations with Hammett σ constants [258]. Electron withdrawing substituents lead to an increase in the absolute magnitude of the coupling constant. (Since correlations with σ^+ constants are worse, the authors [258] consider back-bonding in these compounds unimportant). The coupling constants also fit the more general correlation [258,259] that holds for $R^1R^2R^3\text{SiH}$ compounds:

$$^1J(^{29}\text{Si}-^1\text{H}) = -10.21 \sum \sigma^* - 182.9 \qquad (30)$$

This correlation was used [260] for determination of Taft polar σ^* constants of some X-C_6H_4 groups. The coupling constant can be related to the s-character of the silicon hybrid orbital [258,261], but correction for an extra contribution from the phenyl ring should be made [261].

Chemical shifts of Si-H protons in phenylsilanes such as m,p-Y-C_6H_4-$SiH_{3-n}(CH_3)_n$ have also been linearly correlated with Hammett σ constants [258]. The slopes found in these correlations (2.7-4.1 Hz/σ unit) are considerably smaller than those found in the corresponding dependences for substituted toluenes (12.8 Hz/σ unit) [258]. The solvent reaction field influences SiH_3 proton chemical shifts more than it affects CH_3 shifts in p-tolylsilane [262].

The ^{13}C and ^1H NMR spectra of methyl groups in substituted phenyltrimethylsilanes received considerable attention in connection with the studies of substituent effect propagation into the side-chains of substituted benzenes. The ^{13}C chemical shifts reported by Schaffer et al. [263] for para substituted compounds were subjected to several analyses, although the overall variation of this shift is small (less than 0.5 ppm [263] or 0.8 ppm [196]), and the shifts are substantially affected by solvents [196]. First, the original DSP (σ_I, σ_R) analysis [263] has shown that the substituent effects are transmitted to this carbon about equally by inductive and resonance mechanisms. The dependence is "inverse", i.e., the substituents with electron-withdrawing power cause an increase in the carbon shielding. The DSP analysis [263] and detailed statistical treatment employing a single substituent constant (σ)

[249] have shown that the ability of silicon to transmit substituent effects is similar to that of carbon in analogous t-butylbenzenes. According to another DSP analysis ($\underline{i}$, σ_R) [264], a 1,3-through space interaction between the methyl carbon and the benzene carbon atoms is responsible for the "inverse" character of the correlation. It appears that the ^{13}C-SCS theory of Wolff and Radeglia [265] could account for the observed effects in ^{13}C NMR spectra, since its modification for ^{29}Si shieldings (the σ_E term in the ERW theory) can explain the ^{29}Si chemical shifts in these compounds.

The $^1J(^{13}C-^1H)$ coupling constants in the methyl groups correlate best with Hammett σ values when they are treated by SSP analysis separately in o-, m-, and p-substituted phenyltrimethylsilanes [266], m- and p-substituted phenyldimethylsilanes, and phenylmethylsilanes [258]. The correlations are "normal", i.e., electronegative substituents increase the s-character of the C-H bond and thus produce larger coupling constants. The overall variation in these couplings is very small (1.4 Hz). For that reason, a proper statistical analysis of the analogous correlations that hold for the corresponding compounds of the other group IV elements is not conclusive about which of the elments is a better transmitter of substituent effects [249].

Similar "normal" SSP correlations (σ) were described for methyl proton chemical shifts [251,258,266-268] (the shielding decreases with the increasing electron-withdrawing power of the o-, m-, and p-Y substituent). When the slopes in such correlations were compared among analogous group IVB compounds, the transmitting ability of C, Si, and Ge appeared to be the same, with the transmitting ability of silicon being perhaps somewhat less than those of carbon and germanium [251]. In view of the larger size of the silicon, compared to the carbon atom, this observation had to be ascribed either to the greater polarizability of the Si atom or to $(p-d)_\pi$ interaction [251]. The latter interpretation was favored by Freeburger and Spialter [266] on the basis of observations of $^1J(^{13}C-^1H)$ coupling constants. On the other hand, Nagai et al. [258] argue against this interpretation in phenylmethylsilanes on the basis that $^1J(^{29}Si-^1H)$ couplings fail to correlate with σ^+ constants.

Earlier in this section, the reasons for correlations of phenolic OH proton chemical shifts with Hammett constants were discussed. For similar reasons, excellent correlations exist for silanolic OH proton chemical shifts of substituted phenylsilanetriols dissolved in dimethylsulfoxide (DMSO), dimethylformamide (DMF), and hexamethylphosphoramide (HMPA) [269]. The correlations are "normal" with electron-withdrawing substituents causing deshielding; the effects are about equal in DMSO and DMF, but in HMPA are somewhat larger [269].

Ortho-substituted phenylsilanes have been studied to a much lesser extent. Scholl et al. [51] have noted that o-nitro and -methylphenyltrimethylsilanes have essentially the same ^{29}Si chemical shifts as their para isomers. There was no indication that the same factors determine the shifts of both isomers with no appreciable steric or other effects. Additional small effects were discerned later [270]. The ^{13}C chemical shifts of $Si(CH_3)_3$ groups are always deshielded by ortho positioned methyl, isopropyl, and trimethylsilyl groups. The steric effects on both ^{13}C and ^{29}Si chemical shifts are smaller in trimethylsilyl than in t-butyl-benzenes. The correlations of $^1J(^{13}C-^1H)$ discussed above are of equal quality for the ortho isomers, but no correlation could be found for ^1H methyl proton chemical shifts (for reasons attributed to anisotropy effects) [266].

3.4.3 NMR Spectroscopy of the Connecting Chain

The data of the last group, i.e., ^1H and ^{13}C NMR parameters of disubstituted benzene rings, are not only interesting for the studies of substituent effect propagation, but they are also of analytical importance, especially for distinguishing positional isomers.

The influence of positional isomerism or of molecular symmetry on the appearance of ^1H and ^{13}C NMR spectra of polysubstituted benzenes are well described in the standard texts [84]. If the question of positional isomerism cannot be resolved directly on the basis of the spectral pattern, the numerical values of the NMR parameters must be used.

The $^1H-^1H$ coupling constants are in this respect

somewhat less convenient, since their dependence on the number of intervening bonds is precluded by their dependence on the nature of the substituents. The chemical shifts are to a good approximation additive, and the contributions of various substituents into ortho, meta, and para positions have been tabulated both for ^{1}H [271] and ^{13}C chemical shifts [272]. Since complex spectra of strongly coupled aromatic protons require spectral analysis before they can be interpreted in structural terms, it is usually more convenient to use ^{13}C NMR spectra for this purpose, which (since they are routinely measured with incoherent decoupling of protons) are simple and the chemical shifts can be read out directly. Table XXI gives SCS values for a number of silyl gorups. These values, when combined with the tabulated values [272] for other ring substituents, should give precise predictions of chemical shifts to within ±1 ppm, which would be sufficient to distinguish most of the positional isomers. (Small solvent effects, 0.6 ppm maximum, were observed [273] on the carbon chemical shifts of $p\text{-}CH_3C_6H_4SiCl_3$). The standard deviation for 80 observed chemical shifts was 0.88 ppm. Larger deviations, up to 1.0 ppm, are usual for C-1 substituents bearing carbon atoms [196]. Deviations larger than 2 ppm are indicative of substituent interactions or ring distortions [270].

Systematic trends in the small deviations from additivity were also discerned [274,275] for various series of compounds with one fixed substituent. It was found that better predictions of C-1 chemical shifts can be obtained if the direct additivity is replaced by a proportionality relationship. In such a relationship the SCS values are multiplied by a proportionality constant $\underline{b}$ that can be optimized for a given fixed sustituent in a series [274]. The values of the $\underline{b}$ proportionality constant for C-1 chemical shifts appear related to the reciprocal of the ionization potential of the key atom of the substituent (in this case, the Si) [274] and to the inductive substituent (parameter $\underline{i}$) [276].

The ^{13}C chemical shifts show an acceptable correlation with CNDO/2 calculated electron densities [277]. The aromatic carbon chemical shifts in para substituted phenyltrimethylsilanes [263] were analyzed by the DSP method (σ_I, σ_R^o) [263]. The correlations were good for C-1 and C-2 atoms (i.e., for carbons bearing the silicon atom and ortho to it). The C-1 shifts indicated a high dependence on the π-electron density [263]. This finding is in agreement with the trends established for para carbon atoms in monosubstituted benzenes [132]. It was not possible to distinguish between $(p-d)_\pi$ and π-σ interactions on the basis of ^{13}C chemical shifts [263]. According to the C-4 chemical shift of unsubstituted and para substituted phenyltrimethylsilanes [141,263,278,279], the $Si(CH_3)_3$ group exerts electron-withdrawing effects that are predominantly a π-effect and which outweigh its donating capacity [263]. (Compare the effects of the $CH_2Si(CH_3)_3$ group discussed in Section 3.3.4).

In contrast, it has been concluded from the dependence of $^3J(^1H-^1H)$ on the sum of the electronegativities of the substituents that, unless the para substituent is a strong electron donor, the $Si(CH_3)_3$ group affects the aromatic ring essentially by its inductive effect [280]. When the para substituent Y is a strong electron donor, $(p-d)_\pi$ interaction takes place [217,277,280].

3.5 CONCLUSIONS

Analytical applications of NMR spectroscopy to carbon-functional compounds have been and certainly will remain to be the domain of ^1H and ^{13}C NMR (and perhaps of combinations such as heteronuclear selective decoupling experiments). This analytical role of NMR spectroscopy is widely accepted. Some of the data collected and reviewed here should be helpful in this respect (e.g., ^{13}C SCS values).

On the other hand, it has been seen that ^1H NMR spectroscopy can offer several tools for conformer population analysis. Irrespective of the importance of conformations for understanding the chemistry of carbon-functional organosilicon compounds, this potential of NMR has not yet been exploited.

Interpretation of NMR data in structural terms (electron density and the like) is severely limited by the lack of adequate theory that is sufficiently simple (i.e., clearly related to the concepts of physical organic chemistry) and yet generally reliable and physically sound. In several instances it can be seen that ^{13}C chemical shifts are most promising in this respect. Proton chemical shifts are not likely to furnish much reliable information about bonding situations in carbon-functional compounds. Discussions of ^{29}Si chemical shifts have shown a number of examples of contradictory interpretations of the same experimental data. The ERW theory certainly represents a step forward, but although it yields correct predictions of trends in many cases, it provides little more insight than the quantum chemical method it employs. Obviously, more theoretical efforts are still needed to ensure an unambiguous interpretation of NMR data.

REFERENCES

1. Bažant V., Horák M., Chvalovský V., Schraml J.,: Handbook of Organosilicon Compounds Vol. 1. Advances in Organosilicon Chemistry p. 11, M. Dekker, Inc., New York 1975.
2. Harris R.K., Mann B.E., eds.: NMR and the Periodic Table. Academic Press, New York 1978.
3. Marsmann H.: "^{29}Si-NMR Spectroscopic Results", in NMR Basic Principles and Progress (Diehl P., Fluck E., Kosfeld R., eds.). Vol 17, Springer-Verlag, Berlin 1981.
4. Williams E.A., Cargioli J.D.: "Silicon-29 NMR Spectroscopy", in Annual Reports on NMR Spectroscopy (Webb G.A., ed.) Vol. 9, p. 221. Academic Press, London 1979.
5. Harris R.K.: Chapter 10A in NMR and the Periodic Table (Harris R.K., Mann B.E., eds.). Academic Press, New York 1978.
6. Schraml J., Bellama J.M.: "^{29}Si Nuclear Magnetic Resonance," in Determination of Organic Structures by Physical Methods (Nachod F.C., Zuckerman J.J., Randall E.W., eds.). Vol. 6, p. 203. Academic Press, New York 1976.
7. Noggle J.P., Schirmer R.E.: The Nuclear Overhauser Effect, Chemical Applications. Academic Press, New York 1971.
8. McFarlane W.: Magnetic Multiple Resonance, in Nuclear Magnetic Resonance Spectroscopy of Nuclei Other than Protons (Axenrod T., Webb G.A., eds.). Wiley, New York 1974.
9. Johannesen R.B., Farrar T.C., Brinckman F.E., Coyle T.D.: J. Chem. Phys. 44, 962 (1966).

10. Johannesen R.B., Coyle T.D.: Endeavour 31, 10 (1972).
11. Baker E.B.: J. Chem. Phys. 37, 911 (1962).
12. van den Berghe E.V., van der Kelen G.P.: J. Organometal. Chem. 59, 175 (1973).
13. McFarlane W., Seaby J.M.: J. Chem. Soc., Perkin Trans. 2, 1561 (1972).
14. Pestunovich V.A., Albanov A.I., Larin M.F., Ignateva L.P., Voronkov M.G.: Izv. Akad. Nauk. SSSR, Ser. Khim. 2185 (1978).
15. Pestunovich V.A., Tandura S.N., Shternberg B.Z., Baryshok V.P., Voronkov M.G.: Izv. Akad. Nauk. SSSR, Ser. Khim. 2653 (1978).
16. Holzman G.R., Lauterbur P.C., Anderson J.H., Koth W.: J. Chem. Phys. 25, 172 (1956).
17. Lauterbur P.C., in Determination of Organic Structures by Physical Methods (Nachod F.C., Phillips W.D., eds.). Vol. 2, p. 465. Academic Press, New York 1962.
18. Farrar T.C., Becker E.D.: Pulse and Fourier Transform NMR. Academic Press, New York 1971.
19. Shaw D.: Fourier Transform N.M.R. Spectroscopy. Elsevier, Amsterdam 1976.
20. Levy G.C., Cargioli J.D.: "^{29}Si Fourier Transform NMR," in Nuclear Magnetic Resonance Spectroscopy of Nuclei Other than Protons (Axenrod T., Webb G.A., eds.) p. 251. Wiley (Interscience), New York 1974.
21. Harris R.K., Kimber B.J.: Appl. Spectr. Rev. 10, 117 (1975).
22. Harris R.K., Kimber B.J.: J. Organometal. Chem. 70, 43 (1974).
23. Li S., Johnson D.L., Gladysz J.A., Servis K.L.: J. Organometal. Chem. 166, 317 (1979).
24. Chingas G.C., Garroway A.N., Moniz W.B., Bertrand R.D.: J. Amer. Chem. Soc. 102, 2526 (1980).
25. Linde S.A., Jakobsen H.J., Kimber B.J.: J. Amer. Chem. Soc. 97, 3219 (1975).
26. Maudsley A.A., Muller L., Ernst R.R.: J. Magn. Resonance 28, 463 (1977).
27. Bertrand R.D., Moniz W.B., Garroway A.N., Chingas G.C.: J. Amer. Chem. Soc. 100, 5227 (1978).
28. Murphy P.D., Taki T., Sogabe T., Metzler R., Squires T.G., Gerstein B.C.: J. Amer. Chem. Soc. 101, 4055 (1979).
29. Chingas G.C., Bertrand R.D., Garroway A.N., Moniz W.B.: J. Amer. Chem. Soc. 101, 4058 (1979).
30. Bertrand R.D., Moniz W.B., Garroway A.N., Chingas G.C.: J. Magn. Resonance 32, 465 (1978).

31. Chingas G.C., Garroway A.N., Bertrand R.D., Moniz W.B.: J. Magn. Resonance 35, 283 (1979).
32. Craig R.A., Harris R.K., Morrow R.J.: Org. Magn. Resonance 13, 229 (1980).
33. Morris G.A., Freeman R.: J. Amer. Chem. Soc. 101, 760 (1979).
34. Morris G.A.: J. Amer. Chem. Soc. 102, 428 (1980).
35. Doddrell D.M., Pegg D.T., Brooks W., Bendall M.R.: J. Amer. Chem. Soc. 103, 727 (1981).
36. Bolton P.H.: J. Magn. Resonance 41, 287 (1980).
37. Bolton P.H., James T.L.: J. Amer. Chem. Soc. 102, 1449 (1980).
38. Pachler K.G.R., Wessels P.L.: Org. Magn. Resonance 9, 557 (1977).
39. Pachler K.G.R., Wessels P.L.: J. Magn. Resonance 28, 53 (1977).
40. Pachler K.G.R., Wessels P.L.: J. Magn. Resonance 12, 337 (1973).
41. Jakobsen H.S., Linde S.A., Sørensen S.: J. Magn. Resonance 15, 385 (1974).
42. Harris R.K., Kimber B.J.: J. Magn. Resonance 17, 174 (1975).
43. Sharp K.G., Sutor P.A., Williams E.A., Cargioli J.D., Farrar T.C., Ishibitsu K.: J. Amer. Chem. Soc. 98, 1977 (1976).
44. Bax A., Freeman R., Kempsell S.P.: J. Magn. Resonance 41, 349 (1980).
45. Bax A., Freeman R.: J. Magn. Resonance 41, 507 (1980).
46. Bax A., Freeman R., Kempsell S.P.: J. Amer. Chem. Soc. 102, 4849 (1980).
47. Mägi M., Schraml J.: Unpublished results.
48. Hunter B.K., Reeves L.W.: Can. J. Chem. 46, 1399 (1968).
49. Harris R.K., Kimber B.J.: Org. Magn. Resonance 7, 460 (1975).
50. Levy G.C., Cargioli J.D., Maciel G.E., Natterstad J.J., Whipple E.B., Ruta M.: J. Magn. Resonance 11, 352 (1973).
51. Scholl R.L., Maciel G.E., Musker W.K.: J. Amer. Chem. Soc. 94, 6376 (1972).
52. Ernst C.R., Spialter L., Buell G.R., Wilhite D.L.: J. Amer. Chem. Soc. 96, 5375 (1974).
53. Engelhardt G., Jancke M, Mägi M., Pehk T., Lippmaa E.: J. Organometal. Chem. 28, 293 (1971).
54. Schraml J.: Unpublished results.
55. Bacon M.R., Maciel G.E., Musker W.K., Scholl R.: J. Amer. Chem. Soc. 93, 2537 (1971).

56. Bacon M.R., Maciel G.E.: J. Amer. Chem. Soc. 95, 2413 (1973).
57. Williams E.A., Cargioli J.D., Larochelle R.W.: J. Organometal. Chem. 108, 153 (1976).
58. Pestunovich V.A., Tandura S.N., Voronkov M.G., Engelhardt G., Lippmaa E., Pehk T., Sidorkin V.F., Zeltshan G.I., Baryshok V.P.: Dokl. Akad. Nauk. SSSR 240, 914 (1978).
59. Engelhardt G., Radeglia R., Jancke H., Lippmaa E., Magi M.: Org. Magn. Resonance 5, 561 (1973).
60. Radeglia R.: Z. phys. Chem. (Leipzig) 256, 453 (1975).
61. Schraml J., Nguyen-Duc-Chuy, Chvalovský V., Mägi M., Lippmaa E.: Org. Magn. Resonance 7, 379 (1975).
62. Vongehr M., Marsman M.C.: Z. Naturforsch. 31b, 1423 (1976).
63. Schraml J., Chvalovský V., Mägi M., Lippmaa E.: Coll. Czech. Chem. Commun. 46, 377 (1981).
64. Raynes W.T.: "Theoretical and Physical Aspects of Nuclear Shielding," in Nuclear Magnetic Resonance, A Specialist Periodical Report (Abraham R.J., ed.) p. 2, Vol. 8. The Chemical Society, London 1979.
65. Webb G.A.: "Background Theory of NMR Parameters," in NMR and the Periodic Table (Harris R.K., Mann B.E., eds.) Academic Press, New York 1978.
66. Chapter 1 in Nuclear Magnetic Resonance (A Specialist Periodical Report). The Royal Society of Chemistry, London.
67. Ridard J., Levy B., Millie P.: Mol. Phys. 36, 1025 (1978).
68. Lyubimov V.S., Ionov S.P.: Russ. J. Chem. Phys. 46, 486 (1972).
69. Wolff R., Radeglia R.: Org. Magn. Resonance 9, 64 (1977).
70. Wolff R., Radeglia R.: Z. phys. Chem. (Leipzig) 257, 181 (1976).
71. Wolff R., Radeglia R.: Z. phys. Chem. (Leipzig) 258, 145 (1977).
72. Wolff R., Radeglia R.: Z. phys. Chem. (Leipzig) 261, 726 (1980).
73. Wolff R., Radeglia R., Ponec R.: Org. Magn. Resonance 20, 31 (1982).
74. Jameson C.J., Gutowsky H.S.: J. Chem. Phys. 40, 1714 (1964).
75. Roelandt F.F., van de Vondel P.F., van den Berghe E.V.: J. Organometal. Chem. 94, 377 (1975).

76. Radeglia R., Engelhardt G.: J. Organometal. Chem. 67, C-45 (1974).
77. Horn M., Murrell J.N.: J. Organometal. Chem. 70, 51 (1974).
78. Zeeck E.: Theor. Chim. Acta 35, 301 (1974).
79. Letcher J.H., van Wazer J.R.: "Quantum-Mechanical Theory of ^{31}P NMR CHemical Shifts in Topics," in Phosphorus Chemistry Vol. 5, ^{31}P Nuclear Magnetic Resonance (Grayson M., Griffith E.J., eds.) Interscience, New York 1967.
80. Buckingham A.D.: Can. J. Chem. 38, 300 (1960).
81. Schraml J., Ponec R., Chvalovský V., Engelhardt G., Jancke H., Kriegsmann H., Larin M.F., Pestunovich V.A., Voronkov M.G.: J. Organometal. Chem. 178, 55 (1979).
82. Cantacuzene J., Jantzen R., Tordeux M., Chachaty C.: Org. Magn. Resonance 7, 407 (1975).
83. Kelly D.P., Underwood G.R., Parron P.F.: J. Amer. Chem. Soc. 98, 3106 (1976).
84. Emsley J.W., Feeney J., Sutcliffe L.H.: High Resolution Nuclear Magnetic Resonance Spectroscopy Vols. 1 and 2. Pergamon Press, Oxford 1966.
85. "Nuclear Spin-Spin Coupling," in A Specialist Periodical Report on Nuclear Magnetic Resonance. The Chemical Society, London.
86. Jankowski K., Bélanger J., Söler F., Zamojski A.: Org. Magn. Resonance 12, 544 (1979).
87. Bothner-By A.A.: "Geminal and Vicinal Proton-Proton Coupling Constants in Organic Compounds," in Advances in Magnetic Resonance (Waugh J.S., ed.) Vol 1, p. 195. Academic Press, New York 1965.
88. Kowalewski J.: Progr. NMR Spectroscopy 11, 1 (1978).
89. Bystrov V.F.: Progr. NMR Spectroscopy 10, 41 (1976).
90. Powles J.G., Strange J.H.: Mol. Phys. 5, 329 (1962).
91. Whipple E.B.: J. Magn. Resonance 5, 163 (1971).
92. Pachler K.G.R.: J. Magn. Resonance 8, 183 (1972).
93. Samek Z.: Coll. Czech. Chem. Commun. 43, 3210 (1978).
94. Karplus M.: J. Chem. Phys. 30, 11 (1959).
95. Karplus M.: J. Amer. Chem. Soc. 85, 2870 (1963).
96. Haasnoot C.A.G., de Leeuw F.A.A.M., de Leeuw H.P.M., Altona C.: Org. Magn. Resonance 15, 43 (1981).
97. Jankowski K.: Org. Magn. Resonance 10, 50 (1977).
98. Vorontsova L.G., Bochkov A.F.: Org. Magn. Resonance 6, 654 (1974).
99. Jankowski K., Rabczenko A.: Org. Magn. Resonance 9, 480 (1977).
100. Lambert J.B.: Accounts Chem. Res. 4, 87 (1971).

101. Lambert J.B., Sun H.N.: Org. Magn. Resonance 9, 621 (1977).
102. Durette P.L., Horton D.: Org. Magn. Resonance 3, 417 (1971).
103. Schug J.C., McMahon P.E., Gutowsky H.S.: J. Chem. Phys. 33, 843 (1960).
104. Lin W.C.: J. Chem. Phys. 50, 1890 (1969); 58, 4971 (1973).
105. Carleer R., Anteunis M.J.O.: Org. Magn. Resonance 13, 253 (1980).
106. Ouellette R.J., Baron D., Stolfo J., Rosenblum A., Weber P.: Tetrahedron 28, 2163 (1972).
107. Ouellette R.J., Williams S.H.: J. Amer. Chem. Soc. 93, 466 (1971).
108. Ihrig A.M., Smith S.L.: J. Amer. Chem. Soc. 92, 759 (1970).
109. Bystrov V.F., Cavrilov Yu.D., Solkan V.N.: J. Magn. Resonance 19, 123 (1975).
110. Hansen P.E.: Org. Magn. Resonance 11, 215 (1978).
111. Bystrov V.F., Stepanyants A.U.: J. Mol. Spectroscopy 21, 241 (1966).
112. Barfield M., Spear R.J., Sternhell S.: Chem. Rev. 76, 593 (1976).
113. Saunders J.K., Easton J.W.: "The Nuclear Overhauser Effect," in Determination of Organic Structures by Physical Methods (Nachod F.C., Zuckerman J.J., Randall E.W., eds.) Vol 6, p. 271. Academic Press, New York 1976.
114. Dědina J., Schraml J.: Coll. Czech. Chem. Commun. 42, 3239 (1977).
115. Barry C.D., North A.C.T., Glasel J.A., Williams R.J.P., Xavier A.V.: Nature 232, 236 (1971).
116. Sievers R.E. (Ed.): Nuclear Magnetic Resonance Shift Reagents, Academic Press, New York 1973.
117. Yasterbov V.V., Zhavoronkov I.P., Tyurikov V.A.: Zh. Strukt. Khim. 20, 143 (1979).
118. Emsley S.W., Lindon J.C.: NMR Spectroscopy Using Liquid Crystal Solvents, Pergamon Press, Oxford 1975.
119. Diehl P., Khetrapal C.L.: "NMR Studies of Molecules Oriented in the Nematic Phase of Liquid Crystals," in NMR Basic Principles and Progress (Diehl P., Fluck E., Kosfeld R., eds.) Vol. 1, p. 1. Springer, Heidelberg 1969.
120. Lunazzi L.: "Molecular Structures by NMR in Liquid Crystals," in Determination of Organic Structures by Physical Methods (Nachod F.C., Zuckerman J.J.,

Randall E.W., eds.) Vol. 6, p. 335. Academic Press, New York 1976.
121. Khetrapal C.L., Becker E.D.: J. Magn. Resonance 43, 8 (1981).
122. Fung B.M., Wei I.Y.: J. Amer. Chem. Soc. 92, 1497 (1970).
123. Snyder L.C., Meiboom S.: J. Chem. Phys. 44, 4057 (1966).
124. Ader R., Loewenstein A.: Mol. Phys. 27, 1113 (1974).
125. Ader R., Loewenstein A.: J. Amer. Chem. Soc. 96, 5336 (1974).
126. Diehl P., Kellerhals H.P., Niederberger W.: J. Magn. Resonance 4, 352 (1971).
127. Gil V.M.S., Geraldes C.F.G.C.: in Nuclear Magnetic Resonance Spectroscopy of Nuclei Other than Protons (Axenrod T., Webb G.A., eds.) p. 219. Wiley, New York 1974.
128. Egorochkin A.N., Mironov V.F., Voronkov M.G.: Zh. Strukt. Khim. 7, 450 (1966).
129. Gouverneur P., Nagy O.B., Soumillion J.P.H., Burton T., Bruylants A.: Org. Magn. Resonance 4, 39 (1972).
130. Lazzeretti P., Taddei F.: Org. Magn. Resonance 3, 113 (1971).
131. Del Re G. Pullman B., Yonezawa T.: Biochim. Biophys. Acta 75, 153 (1963).
132. Levy G.C., Nelson G.L.: Carbon-13 Nuclear Magnetic Resonance for Organic Chemists. Wiley, New York 1972.
133. Breitmaier E., Voelter W.: ^{13}C NMR Spectroscopy, Methods and Applications, Verlag Chemie, Weinheim 1974.
134. Wehrli F.W., Wirthlin T.: Interpretation of Carbon-13 NMR Spectra. Heyden, London 1976.
135. Martin G.J., Martin M.L., Odiot S.: Org. Magn. Resonance 7, 2 (1975).
136. Miyasima G., Nishimoto K.: Org. Magn. Resonance 6, 313 (1974).
137. Schraml J., Chvalovský V., Mägi M., Lippmaa E.: Coll. Czech. Chem. Commun. 44, 854 (1979).
138. Schraml J., Včelák J., Engelhardt G., Chvalovský V.: Coll. Czech. Chem. Commun. 41, 3758 (1976).
139. Schraml J., Nguyen-Duc-Chuy, Novák P., Chvalovský V., Mägi M., Lippmaa E.: Coll. Czech. Chem. Commun. 43, 3202 (1978).
140. Kelling H., Zingler G.: private communication.
141. Nguyen-Duc-Chuy, Chvalovský V., Schraml J., Mägi M., Lippmaa E.: Coll. Czech. Chem. Commun. 40, 875 (1975).

142. Schraml J., Včelák J., Chvalovský V.: Coll. Czech. Chem. Commun. 39, 267 (1974).
143. Hartwell G.E., Brown T.L.: J. Amer. Chem. Soc. 88, 4625 (1966).
144. Hartwell G.E., Allenhand A.: J. Amer. Chem. Soc. 93, 4415 (1971).
145. Glockling F., Stobart S.R., Sweeney J.J.: J.C.S. Dalton 2029 (1973).
146. Brady F., Henrick K., Matthews R.W., Gillies D.G.: J. Organometal. Chem. 193, 21 (1980).
147. Sadia A.H., Smith J.D.: J. Organometal. Chem. 153, 253 (1978).
148. Leach J.B., Oates G., Handley J.B., Fung A.P., Onak T.: J.C.S. Dalton 819 (1977).
149. Ungermann C.B., Onak T.: Inorganic Chem. 16, 1428 (1977).
150. Todd L.J., Siedle A.R.: Progr. NMR Spectrosc. 13, 87 (1980).
151. Sonnek G., Reinheckel H.: Z. Chem. 16, 65 (1976).
152. Numata S., Kurosawa H., Okawara R.: J. Organometal. Chem. 70, C21 (1974).
153. Kurosawa H., Numata S., Konishi T., Okawara R.: Bull. Chem. Soc. Jap. 51, 1397 (1978).
154. Schmidbaur H.: Chem. Ber. 97, 270 (1964).
155. Mironov V.F., Gar T.K., Bulvakov A.A., Slobodina V.M., Guntsadze T.P.: Zh. Obshch. Khim. 42, 2010 (1972).
156. Williams K.C.: J. Organometal. Chem. 23, 465 (1970).
157. Collier M.R., Lappert M.F., Pearce R.: J.C.S. Dalton 445 (1973).
158. van Dyke, C.H., Kifer E.W., Gibbon G.A.: Inorg. Chem. 11, 408 (1972).
159. Gynane M.J.S., Lappert M.F., Miles S.J., Carty A.J., Taylor N.J.: J.C.S. Dalton 2009 (1977).
160. Mislow K., Raban M.: Topics in Stereochemistry 1, 1 (1967).
161. Jennings W.B.: Chem. Rev. 75, 307 (1975).
162. Čapka M., Schraml J., Jancke H.: Coll. Czech. Chem. Commun. 43, 3347 (1978).
163. Schraml J., Čapka M., Jancke H.: Coll. Czech. Chem. Commun. 47, 793 (1982).
164. Singh G., Reddy G.S.: J. Org. Chem. 44, 1057 (1979).
165. Dědina J., Schraml J., Chvalovský V.: Coll. Czech. Chem. Commun. 37, 3762 (1972).
166. Brook A.G., Pannell K.H.: J. Organometal Chem. 8, 179 (1967).
167. Dyer J., Lee J.: Spectrochim. Acta 26A, 1045 (1970).

168. Schraml J., Chvalovský V., Mägi M., Lippmaa E.: Coll. Czech. Chem. Commun. 43, 3365 (1978).
169. Bellama J.M., MacDiarmid A.G.: J. Organometal. Chem. 18, 275 (1969).
170. Schmidbaur H., Zimmer B., Kohler F.H., Buchner W.: Z. Naturforsch. 32b, 481 (1977).
171. Organosilicon Compounds (Bažant V., Chvalovský V., Rathouský J., eds.) Vols. 3-9. Publications of the Institute of Chemical Process Fundamentals of the Czechoslovak Academy of Sciences, Prague 1973-1982.
172. Lindman B., Forsén S.: "Chlorine, Bromine and Iodine NMR, Physico-Chemical and Biological Applications," in NMR Basic Principles and Progress (Diehl P., Flucke E., Kosfeld R., eds.). Vol. 12, Springer-Verlag, Berlin 1976.
173. Voronkov M.G., Feshin V.P., Djakov V.M., Romanenko L.S., Baryshok V.P., Sigalov M.V.: Dokl. Akad. Nauk. SSSR 223, 1133 (1975).
174. Ponec R., Chvalovský V.: Coll. Czech. Chem. Commun. 40, 2309 (1975).
175. Liu E.K.S., Lagow R.J.: J. Organometal. Chem. 145, 167 (1978).
176. Bent H.A.: J. Inorg. Nucl. Chem. 19, 43 (1961).
177. Lippmaa E., Mägi M., Engelhardt G., Jancke H., Chvalovský V., Schraml J.: Coll. Czech. Chem. Commun. 39, 1041 (1974).
178. Bellama J.M., Morrison J.A.: J.C.S. Chem. Commun. 985 (1975).
179. Hairston T.J., O'Brien D.H.: J. Organometal. Chem. 23, C41 (1970).
180. Harbordt C.M., O'Brien D.H.: J. Organometal. Chem. 111, 153 (1976).
181. Nagai Y., Kosugi M., Takeuchi K., Migita T.: Tetrahedron 26, 2791 (1970).
182. Schraml J., Chvalovský V., Mägi M., Lippmaa E.: Coll. Czech. Chem. Commun. 42, 306 (1977).
183. Jones R.G., Partington P., Rennie W.J., Roberts R.M.G.: J. Organometal. Chem. 35, 291 (1972).
184. Egorochkin A.N., Khidekel M.L., Razuvaev G.A., Petuchov G.G., Mironov V.F.: Izv. Akad. Nauk. SSSR, Ser. Khim. 1521 (1964).
185. Egorochkin A.N., Burov A.I., Mironov V.F., Gar T.K., Vyazankin N.S.: Izv. Akad. Nauk. SSSR, Ser. Khim. 775 (1969).
186. Egorochkin A.N., Vyazankin N.S., Khorshev S.Ya.: Izv. Akad. Nauk. SSSR, Ser. Khim. 2074 (1971).

187. Rakita P.E., Worsham L.S.: J. Organometal. Chem. 139, 135 (1977).
188. Grishin Yu.K., Sergeyev N.M., Ustynyuk Yu.A.: Org. Magn. Resonance 6, 413 (1974).
189. Yonemoto T.: J. Magn. Resonance 13, 153 (1974).
190. Cartledge F.K., Riedel K.H.: J. Organometal. Chem. 34, 11 (1972).
191. Shapiro B.L., Proulx T.W.: Org. Magn. Resonance 8, 40 (1976).
192. Redl G., Peddle G.J.D.: J. Phys. Chem. 73, 1150 (1969).
193. Rakita P.E., Rothschild B.J.: Chem. Commun. 953 (1971).
194. Rakita P.E., Wright R.: Inorg. Nucl. Chem. Letters 11, 47 (1975).
195. Zetta L., Gatti G.: Org. Magn. Resonance 4, 585 (1972).
196. Schraml J., Chvalovský V., Mägi M., Lippmaa E.: Coll. Czech. Chem. Commun. 40, 897 (1975).
197. Eaborn C.: J. Chem. Soc. 4858 (1956).
198. Bassindale A.R., Eaborn C., Walton D.R.M., Young D.J.: J. Organometal. Chem. 20, 49 (1969).
199. Bassindale A.R., Eaborn C., Walton D.R.M.: J. Organometal. Chem. 21, 91 (1970).
200. Kitching W., Smith A.J., Adcock W., Rizvi S.Q.A.: J. Organometal. Chem. 42, 373 (1972).
201. Adcock W., Cox D.P., Kitching W.: J. Organometal. Chem. 133, 393 (1977).
202. Adcock W., Rizvi S.Q.A., Kitching W.: J. Amer. Chem. Soc. 94, 3657 (1972).
203. Rizvi S.Q.A., Gupta B.D., Adcock W., Doddrell D., Kitching W.: J. Organometal. Chem. 63, 67 (1973).
204. Adcock W., Gupta B.D., Kitching W., Doddrell D., Geckle M.: J. Amer. Chem. Soc. 96, 7360 (1970).
205. Adcock W., Gupta B.D., Kitching W., Doddrell D.: J. Organometal. Chem. 102, 297 (1975).
206. Reynolds W.F., Hamer G.K., Bassindale A.R.: J.C.S. Perkin Trans. II. 1977, 971.
207. Repinskaya I.B., Rezvukhin A.I., Koptyug V.A.: Zh. Org. Khim. 8, 1647 (1972).
208. Sakurai H., Deguchi S., Yamagata M., Morimoto S.I., Kira M., Kumada M.: J. Organometal. Chem. 18, 285 (1969).
209. Hamer G.K., Peat I.R., Reynolds W.F.: Can. J. Chem. 51, 897 (1973).
210. Dayal S.K., Ehrenson S., Taft R.W.: J. Amer. Chem. Soc. 94, 9113 (1972).
211. Limouzin Y., Maire J.C.: J. Organometal. Chem. 39, 255 (1972).

212. Plzák Z., Mares F., Hetflejs J., Schraml J., Papousková Z., Bazant V., Rochow E.G., Chvalovský V.: Coll. Czech. Chem. Commun. 36, 3115 (1971).
213. Vo-Kim-Yen, Papousková Z., Schraml J., Chvalovský V.: Coll. Czech. Chem. Commun. 38, 3167 (1973).
214. Adcock W., Alste J., Rizvi S.Q.A., Aurangzeb M.: J. Amer. Chem. Soc. 98, 1701 (1976).
215. Adcock W., Aldous G.L., Kitching W.: Tetrahedron Letters 3387 (1978).
216. Adcock W., Aldous G.L., Kitching W.: J. Organometal. Chem. 202, 385 (1980).
217. Angelelli J.M., Maire J.C., Vignollet Y.: J. Organometal. Chem. 22, 313 (1970).
218. Taft R.W., Price E., Fox I.R., Lewis I.C., Andersen K.K., Davis G.T.: J. Amer. Chem. Soc. 85, 3146 (1963).
219. Schraml J.: Studies in Nuclear Magnetic Resonance, NSF Technical Report 1966.
220. Fenton D.E., Zuckerman J.J.: Inorg. Chem. 7, 1323 (1968).
221. Kondo Y., Kondo K., Takemoto T., Ikenone T.: Chem. Pharm. Bull. (Tokyo) 14, 1322 (1966).
222. Tribble M.T., Traynham J.G.: "Linear Correlations of Substituent Effects in ^1H, ^{19}F, and ^{13}C Nuclear Magnetic Resonance Spectroscopy," in Advances in Linear Free Energy Relationships (Chapman N.B., Shorter J., eds.), p. 143, Plenum Press, London 1972.
223. Ewing D.F.: "Correlation of NMR Chemical Shifts with Hammett σ Values and Analogous Parameters," in Correlation Analysis in Chemistry, Recent Advances (Chapman N.B., Shorter J., eds.), p. 357. Plenum Press, New York 1978.
224. Godfrey M.: J.C.S. Perkin II, 769 (1977).
225. Hamer G.K., Peat I.R., Reynolds W.F.: Can. J. Chem. 51, 915 (1973).
226. Dawson D.A., Reynolds W.F.: Can. J. Chem. 53, 373 (1975).
227. Bennett J., Delmas M., Maire J.C.: Org. Magn. Resonance 1, 319 (1969).
228. Friebolin H., Rensch R., Wendel H.: Org. Magn. Resonance 8, 287 (1976).
229. Taft R.W., Price E., Fox I.R., Lewis I.C., Andersen K.K., Davis G.T.: J. Amer. Chem. Soc. 85, 709 (1963).
230. Gutowsky H.S., McCall D.W., McGarvey B.R., Meyer L.H.: J. Amer. Chem. Soc. 74, 4809 (1952).
231. Varlamov A.V., Papoušková Z., Pola J., Trška P., Chvalovský V.: Coll. Czech. Chem. Commun. 42, 489 (1977).
232. Taft R.W.Jr.: J. Chem. Phys. 64, 1805 (1960).

233. Taft R.W.Jr.: J. Amer. Chem. Soc. 79, 1045 (1957).
234. Cohen R.B.: PhD Thesis, Pennsylvania State University, 1966, as quoted in ref. [235].
235. Lipowitz J.: J. Amer. Chem. Soc. 94, 1582 (1972).
236. Taft R.W.: private communication, as quoted in ref. [235].
237. Kondratenko N.V., Syrova G.P., Popov V.I., Shejnker Yu.N., Yagupolskij L.M.: Zh. Obshch. Khim. 41, 2056 (1971).
238. Maire J.C., Angelelli J.M.: Bull. Soc. Chim. France 1311 (1969).
239. Emsley J.W., Phillips L.: Progr. NMR Spectr. 7, 1 (1971).
240. Exner O.: "A Critical Compilation of Substituent Constants," in Correlation Analysis in Chemistry, Recent Advances (Chapman N.B., Shorter J., eds.), p. 439, Plenum Press, New York 1978.
241. Yolles S., Woodland J.H.R.: J. Organometal. Chem. 54, 95 (1973).
242. Yolles S., Woodland J.H.R.: J. Organometal. Chem. 93, 297 (1975).
243. Vignollet Y., Maire J.C., Witanowski M.: Chem. Commun. 1187 (1968).
244. Smith A.J., Adcock W., Kitching W.: J. Amer. Chem. Soc. 92, 6140 (1970).
245. Adcock W., Kitching W.: Chem. Commun. 1163 (1970).
246. Adcock W., Rizvi S.Q.A., Kitching W., Smith A.J.: J. Amer. Chem. Soc. 94, 369 (1972).
247. Schraml J., Chuy N.D., Chvalovský V., Mägi M., Lippmaa E.: J. Organometal. Chem. 51, C5 (1973).
248. Yagupolskii L.M., Ilchenko A.Ya., Kondratenko N.B.: Uspekhi Khim. 43, 64 (1974).
249. Pombrik S.I., Kravtsov D.N., Peregudov A.S., Fedin E.I., Nesmeyanov A.N.: J. Organometal. Chem. 131, 355 (1977).
250. Yoder C.H., Tuck R.H., Hess R.E.: J. Amer. Chem. Soc. 91, 539 (1969).
251. Hess R.E., Haas C.K., Kaduk B.A., Schaefer C.D.Jr., Yoder C.H.: Inorganica Chim. Acta 5, 161 (1971).
252. Katritzky A.R., Topsom R.D.: Angew. Chem. Int. Ed. Engl. 9, 87 (1970).
253. Cutress N.C., Katritzky A.R., Eaborn C., Walton D.R.M., Topsom R.D.: J. Organometal. Chem. 43, 131 (1972).
254. Engelhardt G., Mägi M., Lippmaa E.: J. Organometal Chem. 54, 115 (1973).
255. Marsmann H.C.: Chem. Ztg. 96, 456 (1972).
256. Kroth H.J., Schumann H., Kuivila M.G., Schaeffer C.D.Jr., Zuckerman J.J.: J. Amer. Chem. Soc. 97, 1754 (1975).

257. Ernst C.R., Spialter L., Buell G.R., Wilhite D.L.: J. Organometal. Chem. 59, C-13 (1973).
258. Nagai Y., Ohtsuki M., Nakano T., Watanabe H.: J. Organometal. Chem. 35, 81 (1972).
259. Nagai Y., Matsumoto H., Nakano T., Watanabe H.: Bull. Chem. Soc. Japan 45, 2560 (1972).
260. Mileshkevich V.P., Novikova N.F., Bresler L.S., Zajcev N.B., Korol'ko V.V., Nikolaev G.A.: Zh. Obshch. Khim. 47, 2564 (1977).
261. Rastelli A., Pozzoli S.A.: J. Mol. Struct. 18, 463 (1969).
262. Andrianov K.A., Reikhsfeld V.O., Evdokimo A.M.: Izv. Akad. Nauk SSSR, Ser. Khim. 2358 (1972).
263. Schaeffer C.D.Jr., Zuckerman J.J., Yoder C.M.: J. Organometal. Chem. 80, 29 (1974).
264. Inamoto N., Masuda S.: Chem. Letters 177 (1978).
265. Wolff R., Radeglia R.: Z. phys. Chem. (Leipzig) 263, 1105 (1982).
266. Freeburger M.E., Spialter L.: J. Amer. Chem. Soc. 93, 1894 (1971).
267. Sakurai H., Ohtsuru M.: J. Organometal. Chem. 13, 81 (1968).
268. Jaggard J.F.R., Pidcock A.: J. Organometal. Chem. 16, 324 (1969).
269. Michalska Z., Lasocki Z.: Bull. Acad. Polon. Sci., Ser. Sci. Chim. 19, 757 (1971).
270. Schraml J., Chvalovský V., Mägi M., Lippmaa E., Calas R., Dunogues J., Bourgeois P.: J. Organometal. Chem. 120, 41 (1976).
271. Zanger M.: Org. Magn. Resonance 4, 1 (1972).
272. Ewing D.F.: Org. Magn. Resonance 12, 499 (1979).
273. Matsubayashi G., Tanaka T.: Spectrochim. Acta 30A, 869 (1974).
274. Lynch B.M.: Can. J. Chem. 55, 541 (1977).
275. Bromilow J., Brownle R.T.C., Topsom R.D., Taft R.W.: J. Amer. Chem. Soc. 98, 2020 (1976).
276. Inamoto N., Masuda S., Tori K., Yoshimura Y.: Tetrahedron Letters 4547 (1978).
277. Réffy J., Veszprémi T., Hencsei P., Nagy J.: Acta Chim. Acad. Sci. Hung. 96, 95 (1978).
278. Rakita P.E., Worsham L.S.: J. Organometal. Chem. 137, 145 (1977).
279. Maciel G.E., Natterstad J.J.: J. Chem. Phys. 42, 2427 (1965).
280. Vignollet Y., Maire J.C.: J. Organometal. Chem. 17, P43 (1969).

4

THEORETICAL ASPECTS OF BONDING IN ORGANOSILICON CHEMISTRY

Robert Ponec

Institute of Chemical Process Fundamentals
Czechoslovak Academy of Sciences
Prague, Czechoslovakia

4.1 INTRODUCTION

A comparison of the organic chemistry of carbon and of silicon, respectively, will show the close chemical similarity of the two elements in numerous ways [1]. Despite this similarity, differences do appear, and in some cases the differences are so remarkable that they become a dominant feature in the chemistry of a given element.

The difference between the chemistry of carbon and the chemistry of silicon is most frequently attributed to the possibility of valence shell expansion at silicon. This possibility of utilization of d orbitals for bonding by the elements of the second row was originally proposed by Pauling [2] to explain the existence of species with formally expanded covalence, e.g., SiF_6^{2-}, $RSiF_5^-$, SF_6, PF_5, SO_2Cl_2, etc.

Later, d orbital utilization was also suggested for what is termed $(p-d)_\pi$ or $(p-d)_\sigma$ interaction for compounds with normal covalence [3-6]. This simple idea has become a universal tool for the interpretation of anomalies in the properties and structure of organosilicon compounds. Although there is a general acceptance of the concept of $(p-d)_\pi$ interaction, a number of authors have warned that effects regarded as consistent with such an interaction can also be explained without need of d orbital par-

ticipation [7-10]. For example, the participation of
vacant 4s and 4p orbitals of silicon can serve as an
alternative explanation [9]. Similarly, as demonstrated
by Ebsworth, even the expansion of covalency does not
necessarily require d orbital participation [7]. The bonds
formed in this latter case should, however, be multicenter
ones, as is the case with the boranes. Also, hyperconjugation has been recently proposed as an alternative to
explain observed anomalies [11].

The objective consideration of the relative importance of all of these alternatives is a task for modern
valence theory. It is evident that an extraordinary role
in this respect belongs to the development and application
of quantum chemical methods. This chapter will focus on a
discussion and analysis of the factors influencing the
possibility and extent of d orbital participation and of
other mechanisms of intramolecular interaction in light of
modern valence theory.

4.2 VALENCE SHELL EXPANSION FROM THE POINT OF VIEW OF QUANTUM THEORY

4.2.1 The Concept of Orbitals

The principal tenet of quantum mechanics is the concept of the wave function, which permits the calculation
of all observable properties of the molecule. In the case
of systems larger than the hydrogen atom, the wave function
depends on the space and spin coordinates of all electrons
and nuclei in the molecule. Its analytical form cannot be
obtained from solving the Schrödinger equation, since this
equation is not exactly solvable for systems of more than
two particles. In order to obtain at least an approximate
solution, it is necessary to introduce some simplifying
assumptions about the form of the wave function. First is
the so-called Born-Oppenheimer approximation [12] which
permits the separation of nuclear and electronic motions.
Another approximation concerns the electronic part of the
total wave function; it consists of the construction of
the wave function from certain individual one-electron
functions that are termed orbitals. Orbitals are mathematical functions serving to construct the total many-

electron wave function and have no other physical meaning, an elementary fact that should be emphasized since experimental chemists often show a tendency to consider an orbital as something that really exists. This then leads to an overestimation of the role of the d orbitals of silicon in elucidating the molecular structure of organosilicon compounds. The potential absurdity of such conclusions was very nicely expressed by Seyferth [13].

In a quantum chemical description of molecular structure, the orbitals are generally spread over the region of the whole molecule, and in the framework of the LCAO approximation are expressed in the form of a linear combination of atomic orbitals. These atomic orbitals are centered usually, but not necessarily, on the individual nuclei. The concrete form of individual orbitals is obtained from the solution of the Hartree-Fock equations. Such an approach constitutes the basis of all SCF computing methods.

4.2.2 Classification of Atomic Orbitals

Classification of atomic orbitals as s,p,d, etc., is based both on the fact that in the case of the free atom the operator of orbital momentum L commutes with the Hamiltonian H and also that both operators have a common set of eigenfunctions. For the hydrogen atom, for which it is possible to solve the Schrödinger equation exactly, the corresponding eigenfunctions can be expressed in analytic form (Eq. 1).

$$\Phi_{n,\ell,m}(r,\theta,\phi) = R_{n,\ell}(r) \cdot Y_{\ell,m}(\theta,\phi) \qquad (1)$$

In this equation $R_{n,\ell}(r)$ represents the radial and $Y_{\ell,m}(\theta,\phi)$ the angular part of the wave function. The concrete form of the radial function is not too important, since the directional properties of atomic orbitals that are decisive for bonding utilization are determined solely by the angular part of the wave function described by the spherical harmonics $Y_{\ell,m}(\theta,\phi)$. A detailed discussion of the influence of the radial part of the wave function will be presented in Section 4.2.5.

The form of the spherical harmonics is determined by the values of quantum numbers ℓ and m. The s orbitals correspond to the value of the azimuthal quantum number $\ell=0$, the p orbitals correspond to $\ell=1$, the d orbitals correspond to $\ell=2$, etc. In an isolated atom, therefore, the p orbitals represent a set of three and the d orbitals a set of five degenerate functions corresponding to $2\ell+1$ values of the magnetic quantum number m. Since according to quantum mechanics any linear combination of the degenerate eigenfunctions represents the eigenfunction of the same linear operator, it is possible in the case of the p and d orbitals to transform the set of generally complex functions given by Eq. (1) into a set of real functions. The concrete form of these real functions is represented in the case of the d orbitals by Eqs. (2a-2e), where n in d(n) denotes the value of the azimuthal quantum number ℓ.

$$d_{xy} = -\frac{i}{\sqrt{2}} \; (d(2) - d(-2)) \tag{2a}$$

$$d_{xz} = \frac{1}{\sqrt{2}} \; (d(1) + d(-1)) \tag{2b}$$

$$d_{yz} = -\frac{i}{\sqrt{2}} \; (d(1) - d(-1)) \tag{2c}$$

$$d_{x^2-y^2} = \frac{1}{\sqrt{2}} \; (d(2) + d(-2)) \tag{2d}$$

$$d_{z^2} = d(0) \tag{2e}$$

It should be mentioned that these functions do not represent the only possible set of real d orbitals. As can already be seen from their assignments, the d orbitals are represented by the quadratic expressions in the variables x, y, z. For reasons of symmetry, one can construct from these variables six quadratic functions (x^2, y^2, z^2, xy, xz, yz). The set of these six expressions is not, however, linearly independent. The condition of linear independence fulfills only five of these six functions. For a different choice of these functions, different sets of real orthogonal d orbitals exist, as for example, d_{xy}, d_{xz}, d_{yz}, $d_{z^2-x^2}$, d_{y^2}, or d_{xy}, d_{xz}, d_{yz}, $d_{z^2-y^2}$, d_{x^2}, etc. These "nonstandard" d functions can be used equally well as a basis set of atomic orbitals in place of the known "standard" set of functions described by Eqs. (2a-2e). The transformation between the "nonstandard" and "standard" d functions is described by Eqs. (3a-3d).

$$d_{z^2-x^2} = \frac{\sqrt{3}}{2} d_{z^2} - \frac{1}{2} d_{x^2-y^2} \qquad (3a)$$

$$d_{z^2-y^2} = \frac{\sqrt{3}}{2} d_{z^2} + \frac{1}{2} d_{x^2-y^2} \qquad (3b)$$

$$d_{x^2} = \frac{\sqrt{3}}{2} d_{x^2-y^2} - \frac{1}{2} d_{z^2} \qquad (3c)$$

$$d_{y^2} = -\frac{\sqrt{3}}{2} d_{x^2-y^2} - \frac{1}{2} d_{z^2} \qquad (3d)$$

The "nonstandard" d functions can be encountered, for example, during the application of different symmetry operations on the standard set. For example, the orbital $d_{x^2-y^2}$ transforms itself after rotation around the x axis by 90° to the nonstandard orbital $d_{z^2-x^2}$ (Fig. 1).

This circumstance documents the fact that in contrast to the p orbitals the d orbitals exhibit more complicated transformation properties. A detailed discussion of the transformation properties of d orbitals will be the subject of the following section.

4.2.3 Transformation Properties of d Orbitals

When interpreting molecular electronic structure, one frequently encounters the fact that many current charac-

Fig. 1. Transformation of standard $d_{x^2-y^2}$ orbital to nonstandard $d_{z^2-x^2}$ orbital.

teristics or concepts depend on the actual choice of a coordinate system. In this respect the concept of hybridization may serve as a typical example, since the form of hybrid orbitals depends on the orientation of the molecule in the coordinate system. A quantitative description of this dependence requires a knowledge of the transformation properties of orbitals. These properties can be most simply demonstrated on p orbitals, since the individual p_x, p_y, p_z orbitals transform exactly like ordinary vectors. It means that a given combination of p orbitals ($a_x p_x + a_y p_y + a_z p_z$) is converted after the application of geometry transformation into another combination of original p orbitals ($a_x' p_x + a_y' p_y + a_z' p_z$). Expressed mathematically, it means that the original coordinates a_x, a_y, a_z transform themselves into new coordinates a_x', a_y', a_z', and this transformation is described by the 3x3 matrix, τ_p (Eq. 4).

$$(a_a', a_y', a_z') = (a_x, a_y, a_z) \cdot \tau_p \qquad (4)$$

The form of the τ_p matrix is especially simple in cases when the geometric transformation corresponds to a rotation around some coordinate axis. Thus, for example, the matrix τ_p^x describing the rotation around the x axis by an angle α is described by Eq. (5).

$$\tau_p^x \equiv \begin{pmatrix} 1 & 0 & 0 \\ 0 & \cos\alpha & \sin\alpha \\ 0 & -\sin\alpha & \cos\alpha \end{pmatrix} \qquad (5)$$

Analogous expressions govern the transformation matrices τ_p^y and τ_p^z.

The situation is, however, much more complicated for d orbitals, where it is possible that a given linear combination of the standard basis set orbitals d_{xy}, d_{xz}, d_{yz}, $d_{x^2-y^2}$, d_{z^2} may transform itself into a combination also containing "nonstandard" d functions d_{x^2}, d_{y^2}, etc., that do not appear in the original basis. To illustrate such a situation, the transformation properties of d orbitals can be analyzed in more detail. As an example of transformation, the rotation around some axis by an angle ω can be chosen. From geometry it follows that such a rotation can be described by subsequent rotations around the coordinate axes x, y, z. Discussion will therefore be restricted to the determination of the transformation matrices τ_d^x, τ_d^y, τ_d^z. Because of the five fold degeneracy of d orbitals, these matrices have a 5×5 dimension and their concrete form can be determined on the basis of the following considerations. For sake of brevity, only the derivation of the τ_d^x matrix will be discussed.

On the basis of Eq. (5), which describes the transformation properties of pure x, y, and z vectors for rotation around the x axis, the following expressions describe the transformations of quadratic terms (Eqs. 6a-6e).

$$x'y' = xy \cos \phi + xz \sin \phi \qquad (6a)$$

$$x'z' = -xy \sin \phi + xz \cos \phi \qquad (6b)$$

$$y'z' = \frac{1}{2}(z^2-y^2) \sin 2\phi + yz \cos 2\phi \qquad (6c)$$

$$\frac{1}{2}(x'^2-y'^2) = \frac{1}{2}(x^2-y^2) \cos^2 \phi - \frac{1}{2}(z^2-x^2) \sin^2 \phi - \frac{1}{2} yz \sin 2\phi \qquad (6d)$$

$$z'^2 = y^2 \sin^2 \phi + z^2 \cos^2 \phi - yz \sin 2\phi \qquad (6e)$$

The quadratic expressions in these equations represent just the angular part of real d orbitals (a numerical factor of 1/2 appears for the $d_{x^2-y^2}$, $d_{z^2-x^2}$, $d_{z^2-y^2}$ orbitals). Consequently, one may directly substitute $xy \rightarrow d_{xy}$, $xz \rightarrow d_{xz}$, $z^2 \rightarrow d_{z^2}$, etc. The above equations thus describe the required transformation matrix. From Eqs. (6d) and (6e) one can, however, see that the orbital $d_{x'^2-y'^2}$ is expressed in the form of a combination that also contains the $d_{z^2-x^2}$ orbital, which unfortunately is not a member of the "standard" set of d orbitals. A similar situation also appears for the $d_{z'^2}$ orbital. This inconvenience can be removed by substituting the expressions from Eq. (3) for "nonstandard" functions. After this manipulation, one finally obtains the transformation matrix τ_d^x, the actual form of which is given in Eq. (7b).

$$(a'_{xz}, a'_{yz}, a'_{xy}, a'_{x^2-y^2}, a'_{z^2}) = (a_{xz}, a_{yz}, a_{xy}, a_{x^2-y^2}, a_{z^2}) \cdot \tau_d \qquad (7a)$$

$$\tau_d^x \equiv \begin{pmatrix} \cos\phi & 0 & -\sin\phi & 0 & 0 \\ 0 & \cos 2\phi & 0 & \frac{1}{2}\sin 2\phi & \frac{\sqrt{3}}{2}\sin 2\phi \\ \sin\phi & 0 & \cos\phi & 0 & 0 \\ 0 & -\frac{1}{2}\sin 2\phi & 0 & \cos^2\phi + \frac{1}{2}\sin^2\phi & -\frac{\sqrt{3}}{2}\sin^2\phi \\ 0 & -\frac{\sqrt{3}}{2}\sin 2\phi & 0 & -\frac{\sqrt{3}}{2}\sin^2\phi & \cos^2\phi - \frac{1}{2}\sin^2\phi \end{pmatrix} \quad (7b)$$

The remaining transformation matrices can be similarly derived; their final forms are given by Eqs. (8a) and (8b).

$$\tau_d^y \equiv \begin{pmatrix} \cos 2\psi & 0 & 0 & -\frac{1}{2}\sin 2\psi & \frac{\sqrt{3}}{2}\sin 2\psi \\ 0 & \cos\psi & -\sin\psi & 0 & 0 \\ 0 & \sin\psi & \cos\psi & 0 & 0 \\ \frac{1}{2}\sin 2\psi & 0 & 0 & \cos^2\psi + \frac{1}{2}\sin^2\psi & \frac{\sqrt{3}}{2}\sin^2\psi \\ -\frac{\sqrt{3}}{2}\sin 2\psi & 0 & 0 & \frac{\sqrt{3}}{2}\sin^2\psi & \cos^2\psi - \frac{1}{2}\sin^2\psi \end{pmatrix} \quad (8a)$$

$$\tau_d^z = \begin{pmatrix} \cos\theta & \sin\theta & 0 & 0 & 0 \\ -\sin\theta & \cos\theta & 0 & 0 & 0 \\ 0 & 0 & \cos 2\theta & -\sin 2\theta & 0 \\ 0 & 0 & \sin 2\theta & \cos 2\theta & 0 \\ 0 & 0 & 0 & 0 & 1 \end{pmatrix} \qquad (8b)$$

4.2.4 d Orbitals and Hybridization

The concept of hybridization has proved to be very useful for the qualitative elucidation of molecular electronic structure. In an effort to account for the directional properties of bonds, as well as to satisfy the requirement of the additivity of molecular properties, Pauling [14] showed that by using a properly chosen combination of orbitals on each atom, it is possible to form an equivalent basis of orbitals that are called hybrid orbitals. The bonds formed from these orbitals are localized and mutually independent. It should be realized, however, that hybridization does not correspond to a physical reality but is a purely mathematical operation, the only advantage of which is to retain the classical picture of localized chemical bonds for a description of molecular structure.

The possibility of bonding utilization of different atomic orbitals in forming hybrid bonds is completely determined by the symmetry of the molecule. This fact makes it possible to employ the formalism of group theory for the study of molecular structure. The possible types of hybridization were analyzed by Kimball [15] and Jaffé [5]. They showed that tetrahedral geometry is consistent with two types of hybrid orbitals, sp^3 and sd^3. The specific form of these hybrid orbitals in the coordinate system given in Fig. 2 is described by Eqs. (9) and (10).

$$h_1 = \tfrac{1}{2} s + \tfrac{1}{2} p_x + \tfrac{1}{2} p_y + \tfrac{1}{2} p_z \qquad (9a)$$

$$h_2 = \tfrac{1}{2} s - \tfrac{1}{2} p_x - \tfrac{1}{2} p_y + \tfrac{1}{2} p_z \qquad (9b)$$

$$h_3 = \tfrac{1}{2} s + \tfrac{1}{2} p_x - \tfrac{1}{2} p_y - \tfrac{1}{2} p_z \qquad (9c)$$

$$h_4 = \tfrac{1}{2} s - \tfrac{1}{2} p_x + \tfrac{1}{2} p_y - \tfrac{1}{2} p_z \qquad (9d)$$

$$t_1 = \tfrac{1}{2} s + \tfrac{1}{2} d_{xy} + \tfrac{1}{2} d_{yz} + \tfrac{1}{2} d_{xz} \qquad (10a)$$

$$t_2 = \tfrac{1}{2} s + \tfrac{1}{2} d_{xy} - \tfrac{1}{2} d_{yz} - \tfrac{1}{2} d_{xz} \qquad (10b)$$

$$t_3 = \tfrac{1}{2} s - \tfrac{1}{2} d_{xy} - \tfrac{1}{2} d_{yz} + \tfrac{1}{2} d_{xz} \qquad (10c)$$

$$t_4 = \tfrac{1}{2} s - \tfrac{1}{2} d_{xy} + \tfrac{1}{2} d_{yz} - \tfrac{1}{2} d_{xz} \qquad (10d)$$

In the case of a different orientation of the coordinate system, these expressions transform. The form of the hybrid orbitals can be found with the aid of transformation matrices derived in the previous section.

The fact that tetrahedral geometry is equally well described by sp^3 and sd^3 hybrid orbitals suggests that the

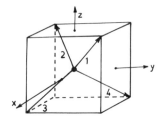

Fig. 2. Coordinate System for Construction of Hybrid Orbitals.

THEORETICAL ASPECTS OF BONDING

most general form of hybrid orbitals is given by a linear combination expressed by Eq. (11) (orbitals h_i and t_i are orthogonal).

$$H_i = a\,h_i + (1-a^2)^{1/2} \cdot t_i \tag{11}$$

Parameter a in this equation measures the relative contribution of sp^3 and sd^3 hybrid orbitals to the total hybrid. The determination of this coefficient thus apparently opens the possibility of a characterization of the extent of d orbital participation. The situation is, however, much more complex. When only s and p valence orbitals contribute to the hybridization, the form of the hybrid orbitals is uniquely determined by the molecular geometry. In the general case described by Eq. (11), which includes the possibility of valence shell expansion, it is necessary to use some external criterion to determine coefficient a. Since several possibilities exist for choosing such a criterion, it is evident that the extent of d orbital participation cannot be determined unambiguously.

To illustrate this conclusion one can analyze in greater detail the above procedure of construction of hybrid orbitals for the silane molecule. As a condition for the optimalization of parameter a, let us choose initially the criterion of maximum overlap [16,17]. This criterion is based on the concept that the strength of the chemical bond is proportional to the values of the overlap integrals between the corresponding orbitals forming the bond. Each of the Si-H bonds is formed by a linear combination of the general hybrid orbital H on the silicon atom with the corresponding 1s orbital of hydrogen. The overlap integral Σ between these orbitals can be expressed in the form of Eq. (12).

$$\Sigma = a\,S_{hs} + \sqrt{1 - a^2}\,S_{ts} \tag{12}$$

where S_{hs} and S_{ts} are individual overlap integrals of pure sp^3 and sd^3 hybrid orbitals of silicon with the 1s orbital of hydrogen. The condition of the maximalization of the overlap integral (Eq. 13) leads to the following form of the general hybrid orbital [18], as shown in Eq. (14).

$$\frac{\partial \Sigma}{\partial a} = 0 \tag{13}$$

$$H = 0.757\ h + 0.653\ t \tag{14}$$

This equation suggests a relatively high contribution of the d orbitals. This conclusion can, however, immediately be questioned. First of all, the extent of d orbital participation depends on the relative values of overlap integrals S_{hs} and S_{ts}, and these quantities are influenced by the values of the exponents in the Slater orbitals. In the standard CNDO/2 method the exponents of both the p and the d orbitals are set equal, which is probably not entirely realistic because of the possible greater diffuseness of d orbitals. A more fundamental objection against such a procedure questions even the physical foundations of the criterion of maximum overlap.

The physically superior method of the optimization of parameter a seems to be based on the following energetic considerations. If the orbital of the Si-H bond is formed by the combination of the general hybrid orbital H with the 1s orbital of hydrogen, the valence state of the silicon can be described as a linear combination of pure sp^3 and sd^3 valence states, and one must assess their relative contributions. The solution of this task leads to the system of secular equations given below, Eq. (15).

$$C_{hh}(\alpha_{hh}-E) + C_{ht}\beta_{th} + C_{hs}\beta_{hs} = 0$$
$$C_{th}\beta_{th} + C_{tt}(\alpha_{tt}-E) + C_{ts}\beta_{ts} = 0 \tag{15}$$
$$C_{sh}\beta_{sh} + C_{st}\beta_{st} + C_{ss}(\alpha_{ss}-E) = 0$$

In order for such equations to be uniquely solvable the determinant of this equation must be equal to zero (Eq. 16).

$$\begin{vmatrix} (\alpha_{hh}-E) & \beta_{ht} & \beta_{hs} \\ \beta_{th} & (\alpha_{tt}-E) & \beta_{ts} \\ \beta_{sh} & \beta_{st} & (\alpha_{ss}-E) \end{vmatrix} = 0 \tag{16}$$

To solve the system of secular equations one needs to know the energies of the sp^3 and sd^3 valence states of silicon. Simple estimates based on the CNDO approximation show that

THEORETICAL ASPECTS OF BONDING

the sp^3 valence state lies roughly 5.9 eV and the sd^3 valence state 17.3 eV above the s^2p^2 ground state configuration [18]. The energetic factors thus act against the involvement of d orbitals in bonding.

The problems arising from the nonunique character of the criterion of the optimalization of the hybridization parameter clearly demonstrate that the concept of hybridization must be regarded as only qualitative, since it cannot be defined unequivocally when working with a larger than minimal basis set of valence orbitals. This fundamental failure thus crucially restricts the applicability of the hybridization concept to estimate the extent of d orbital participation.

4.2.5 Other Factors Influencing the Utilization of d Orbitals in Bonding

The discussion of d orbital participation has been restricted thus far only to directional or angular properties of d orbitals. For concrete quantum chemical calculations it is necessary, however, to specify also the radial part of the wave function. In contemporary quantum chemistry several types of orbitals that differ in their radial part are used. Well known, for example, are the hydrogen-like orbitals of which the radial part is expressed in terms of the Lagguere polynomials. More frequent, however, are the Slater orbitals [19] with the radial part given by the empirical formula expressed in Eq. (17).

$$R_n^S(r) \sim r^{n-1} \cdot e^{-\alpha r} \tag{17}$$

and recently the Gaussian [20] type orbitals described by Eq. 18.

$$R_n^G(r) \sim r^{n-1} \cdot e^{-\alpha r^2} \tag{18}$$

The exponent α appearing in these equations expresses the diffuseness of the orbitals, i.e., it characterizes the distance from the nucleus where the maximum of the radial function is placed. Slater proposed simple empirical rules for the determination of the exponent [19]. On the basis of these rules Craig [3] has shown that d orbitals are too diffuse, which prohibits their efficient participation in bonding. As the simple example demonstrates, the maximum of the radial part of the 3d orbital of silicon with the configuration $3s3p^23d$ is placed roughly at a distance of 9 a.u. (~0.45 nm) from the nucleus [18], which of course exceeds the common range of internuclear distances between bonded atoms. Craig also showed that the extent of d orbital participation may increase in compounds containing strongly electronegative ligands [21], since as a consequence of partial positive charge on an atom of the second row, some effective contraction of d orbitals occurs. All these conclusions based on the diffuseness or contraction of d orbitals are valid, however, only if Slater's rules for the calculation of the orbital exponent are reliable.

As was demonstrated by the results of some ab initio SCF studies, in which the orbital exponent α was determined as a variational parameter from the condition of the minimalization of molecular energy, the values of such exponents suggest that the original ideas on the diffuseness of d orbitals should be considerably modified. For example, Lipscomb [22] reports an optimalized value of exponent $\alpha_d^{Si} = 1.3$ obtained from ab initio calculations of the SiH_4 molecule in the basis of Slater orbitals. This corresponds to a value of the maximum of the radial wave function $r_{max} = 2.3$ a.u., which is much more realistic than the original value of 9 a.u. resulting from Slater's rules. Siegbahn and Roos [23], who determined the optimalized orbital exponent for Gaussian orbitals, also report very similar conclusions. Their value of $\alpha_a^{Si} = 0.3$ corresponds roughly to $r_{max} = 2.2$ a.u., which is entirely comparable to the previous value of Lipscomb and also to the value of r_{max} for 3p orbitals (2.14 a.u.). The value of this exponent does not change significantly within a given row of the periodical table. Pople et al. report that the mean value of the Gaussian orbital exponent for the elements of the second period (Al-Cl) is equal roughly to 0.39 [24]. On the basis of these results it is evident that the restricting influence of the diffuseness of d orbitals that is frequently stated in the literature is probably strongly overestimated.

Another factor influencing the extent of d orbital participation is represented by the energetic requirements connected with the magnitude of the excitation energy

THEORETICAL ASPECTS OF BONDING   247

necessary to populate the vacant d orbitals. The reported
estimates of this excitation energy show that, for
example, the s^2pd configuration of the silicon atom lies
only 6.6 eV above the s^2p^2 ground state configuration
[25]. The authors interpret this result as indicating the
strong involvement of d orbitals in bonding. It could be
argued, however, that the described estimates do not
correspond adequately to the real situation. The problem
lies in the fact that the s^2pd configuration does not
describe the configuration of the valence state of the
quadrivalent silicon. The valence state for a tetrahedrally bonded atom can be characterized either by
sp^3 and sd^3 configurations. The relative contribution of
sp^3 and sd^3 valence states can be determined, as shown in
the previous section, by the solution of the secular determinant (Eq. 16). Energetic factors connected with the
high energy of the sd^3 valence state [18] then act rather
strongly against d orbital participation.

From the theoretical point of view the inclusion of d
orbitals in bonding is also possible for carbon, and it was
in fact proposed by Gillespie [26] to rationalize the
transition state structure in S_N2 substitutions. The considerable energetic requirements connected with the excitation to the sd^3 valence state again renders such
participation negligible.

4.2.6 The Variational Principle and d Orbital
Participation

In the framework of an LCAO approximation the individual molecular orbitals ψ are expressed in the form of a
linear combination of atomic orbitals (Eq. 19).

$$\psi_i = \sum_\mu c_{\mu i} \chi_\mu \qquad (19)$$

The atomic orbitals are usually centered on the
corresponding atoms. This condition is not important,
however, and equally good LCAO expansions can be performed
in the basis of orbitals centered at one atom or even outside the atoms. Such one center calculations were performed for the SiH_4 molecule even on the ab initio level,
and their results are comparable with calculations based on
ordinary LCAO expansion [27,28]. The actual form of the

molecular orbitals is obtained from the solution of
Roothaan equations [29] that are based on the so-called
variational principle. This principle warrants the
finding of the best possible approximate wave function for
a given basis of atomic orbitals. The accuracy with which
this approximate function approaches the exact solution of
the Schrödinger equation depends only on the extent of the
AO basis set. The greater this extent, the greater the
flexibility of the approximate wave function to describe
the real distribution of the electronic density in a molecule. It would, of course, be ideal to work with a
complete, i.e., infinite basis set, but practical mathematical problems require the basis set to be reduced to some
finite extent. In quantum chemical calculations of molecular structure on the ab initio level, one thus encounters
the problem of finding a reasonable compromise between the
size of the basis set and the desired accuracy of the
calculations. In the field of organosilicon chemistry,
however, there is another specific problem, viz., whether
it is necessary to include d orbitals at the silicon atom
for a sufficiently reliable description of the wave function.

From the point of view of the variational principle,
any enlargement of the basis set leads automatically to an
increased accuracy of the wave function. Regarded formally, the variational principle requires the inclusion of
d orbitals into the basis set. However, the situation is
more complex. First of all, it is important whether the
increased accuracy of the wave function substantially
changes the distribution of the electronic density in a
molecule. This cannot be checked, of course, without
detailed calculations. The problem of determination of
the optimum basis set for the silicon atom has been
treated in recent years by several authors [30-33]. The
conclusion of all of these studies is that the most important condition for the correct description of the wave
function is sufficient flexibility of the basis set of s
and p orbitals only. The additional inclusion of d orbitals does not change the overall bonding situation dramatically. The d orbitals act therefore as polarization
functions rather than as true valence shell functions. To
illustrate this conclusion, a recent study describing the
theoretical calculation of Si-O bond lengths in siloxanes
may be used [34]. The marked shortening of the experimental Si-O bond lengths over the value given by the sum of
covalent radii, as well as the large Si-O-Si valence

THEORETICAL ASPECTS OF BONDING

angle, was regarded as clear proof of d orbital participation. Nonempirical ab initio calculations show, however, that a quantitatively correct description of the molecular structure can be obtained even without the inclusion of the d orbitals of silicon. In light of these quantitative results it is evident that the original ideas about the necessity of d orbital participation are very probably incorrect.

Besides these numerically supported arguments, there are further factors that suggest that the concept of $(p-d)_\pi$ and $(p-d)_\sigma$ interaction cannot be given any real physical meaning. This concept is based on the idea of the interaction of mutually overlapping atomic orbitals centered on individual atoms, and the possibilities of such interaction are given by the symmetry of the molecular structure. Such symmetry considerations lie in the basis of the criteria of Sabin and Rodwell [35-37] that evaluate the conditions for the necessity of involvement of d orbitals into the basis set. Another typical example documenting the applicability of symmetry arguments concerns the discussions of the number of π bonds that can be formed by a tetrahedrally bonded silicon atom [15]. As mentioned above, however, the form of LCAO expansion is not restricted only to orbitals centered on individual atoms, but the same electronic density distribution results also from one center calculations. Since this center need not coincide with the position of the silicon atom in a molecule, it is evident that no physical meaning can be attributed either to $(p-d)_\pi$ interaction or to the contributions of d orbitals specifically at silicon. Such conclusions are not restricted only to one-center calculations, but as demonstrated by Coulson [39], they hold also for ordinary LCAO expansion, since the contributions of d orbitals that necessarily appear in the wave function as a consequence of the variational principle cannot be interpreted in the sense of a "weighting" of the extent of d orbital participation. Irrespective of the type of the LCAO expansion, sufficient flexibility of the basis set is thus the only decisive factor in the variational calculations. All subsequent discussions interpreting molecular structure in terms of intramolecular interaction mechanisms represent only a more-or-less convenient means of describing the observed phenomena. In any case, such descriptions cannot be regarded as proof of a real physical existence of these interaction mechanisms. Such conclusions apply not only to the concept of $(p-d)_\pi$ interaction, but also to other possible interaction mechanisms, e.g., hyperconjugation.

4.3 HYPERCONJUGATION

4.3.1 General Introduction

The concept of hyperconjugation was introduced into chemistry in the early forties by Mulliken [40,41], who proposed that the structure of aliphatic radicals could be considered as the resonance hybrid of several mesomeric structures. From the concept of resonance stabilization it follows that there is no fundamental difference between classical conjugation and hyperconjugation except in the magnitude of the effects that they cause. These effects manifest themselves in the form of deviations from the rules of additivity for different molecular properties such as bond lengths (additivity of covalent radii), heats of formation (additivity of bond energies), etc. Whereas for aromatic or classically conjugated systems these deviations are clearly expressed, for aliphatic or nonconjugated molecules they are frequently hindered by experimental errors in the relevant data. This problem led to the concept of hyperconjugation being the subject of nummerous discussions and conflicting opinions. A critical review of this situation can be found in Dewar's classic book [42]. Recently, a new revival of the hyperconjugation concept has arisen from organosilicon chemistry. A critical examination of some newer works documenting the increased role of hyperconjugation in organosilicon chemistry is presented in a subsequent section. The aim of this review is not to give a complete survey of all such works but rather to show some typical examples from different fields of organosilicon chemistry.

4.3.2 Hyperconjugation in Organosilicon Chemistry

For a long time the only generally accepted theory of the structure-reactivity relationships in organosilicon chemistry was based on the concept of $(p-d)_\pi$ or $(p-d)_\sigma$ interaction. The first attempt at an alternative explanation of observed phenomena was the work of Nesmeyanov, who introduced the concept of $\sigma-\sigma$ conjugation [43]. Except

THEORETICAL ASPECTS OF BONDING 251

for the work of several Russian authors who used this idea for the interpretation of anomalous reactivity and spectral behavior of allyl- and benzylsilanes [44-46], the original proposal by Nesmeyanov remained unnoticed until the early seventies, when Pitt [47] proved on the basis of semiempirical CNDO calculations of the phenylsilane molecule that a qualitative understanding of the electron accepting properties of silyl substituents does not necessitate the inclusion of silicon d orbitals into the basis set. These results demonstrate clearly that regardless of possible objections concerning the semiempirical nature of the CNDO method, there must be some other mechanism of intermolecular interaction besides $(p-d)_\pi$ bonding in order to account for the -M effect of silyl substituents.

Important experimental support for the concept of hyperconjugation was provided initially by photoelectron spectroscopy. A large number of papers, especially from the laboratories of Bock [48-55] and Schweig [56-59] appeared in the early seventies. In these papers the usefulness of hyperconjugation for the interpretation of photoelectron spectra was clearly demonstrated. In some cases where PE spectroscopy was unavailable, similar conclusions were also provided by charge-transfer (CT) spectra [60-64]. In one of these works considering the interpretation of CT spectra of substituted silafluorenes [64], it was possible to exclude d orbital participation on the basis of simple symmetry arguments. The interpretation of these CT spectra was based on the simple HMO method with a proper model of the SiX_3 or SiX_2 group [63], allowing the explicit inclusion of hyperconjugation.

In addition to the quantitiave interpretation of CT and PE spectra, the hyperconjugation concept can be supported by numerous qualitative arguments based on steric considerations. As demonstrated on the basis of simple correlation diagrams by Hoffmann et al. [65], hyperconjugation is restricted to those cases where the steric arrangement permits an effective interaction of σ and π orbitals. This demonstration accounts for the stereospecificity of electronic effects of silyl substituents in carbon-functional derivatives. Thus, for example, a strong enhancement of the $I_{C=C}$ intensity of the Raman line is observed in compounds A and B,

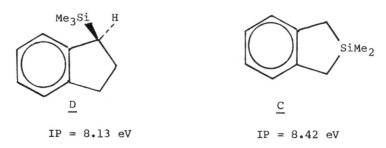

I_{CC} = 780 I_{CC} = 1020 I_{CC} = 345

where the σ orbital of the Si-C bond is perpendicular to the π electron skeleton of the rest of the molecule, in contrast to the planar compound C where a similar interaction is impossible [66,67]. Similarly, the remarkable decrease in the excitation energy of the first CT band in substituted derivatives of cyclopentenylsilanes in comparison with planar silacyclopentenes is also the consequence of σ-π interactions [62]. The same factor is also responsible for the decrease of the ionization potential of compound D in comparison with the planar silaindene, C[68].

D   C

IP = 8.13 eV   IP = 8.42 eV

In the above mentioned examples the hyperconjugation concept was used directly by the authors to explain the observed effects. There are, however, many examples explained originally in term of $(p-d)_\pi$ interaction that can be equally well reinterpreted using the hyperconjugaton concept. To illustrate such a situation we may use the protodesilylation of substituted benzenes, where the low value of the Hammett reaction constant ρ was interpreted by Eaborn in terms of $(p-d)_\pi$ stabilization of the ground state of the phenyltrimethylsilane molecule [69]. Such an interpretation is not, however, the only one possible, and the same conclusions may be obtained by invoking the hyperconjugational stabilization of the benzenonium ion in the transition state. Similarly, the unusually large increase in the rate of protodemetallation in the R_3MPh systems (Table I), which was interpreted as a change of mechanism [70], can be explained by the σ-π interaction of the C_{ar}-M bond with the aromatic nucleus in the transition state, as shown by Barwin [71]. Such an interpretation is also supported by the correlation of the rate of protodemetallation with the excitation energies of CT complexes of parent molecules with TCNE [71].

The concept of hyperconjugation can also be used to interpret the spectral changes considered by Goodman as a test of d orbital participation [72]. In this test it is only important that the silicon enters into the conjugation with one unoccupied orbital of π symmetry, and it is not possible to distinguish on the basis of experimental results alone whether this orbital is the 3d, 4p, or even the σ_π^* antibonding orbital of the SiX bond that is of proper symmetry to interact with the rest of the molecule. Such an orbital would similarly modify the qualitative correlation diagrams proposed by West and Harnish [73,74] for the interpretation of the large bathochromic shift of n-π* transitions in α-silylketones. As the quantum chemical calculations by Agolini [75] and Bock [76] document, the quantitative interpretation of the UV spectra of α-silylketones does not require the inclusion of d orbitals at silicon. The interaction of π orbitals with the σ_π^* orbitals of the SiX bonds in phenylsilanes ($PhSiX_3$) may be used to account for the electron accepting behavior of the SiX_3 group, which manifests itself in the values of the experimental dipole moments [77] (negative hyperconjugation). Also, the variability of the electronic effects of silyl substituents induced by different physicochemical factors such as protonation [78] of the molecule or the change in the nature of the attacking reagent

Table 1. Relative reactivities of R_3MPh molecules in proto-demetallation reactions ($R = CH_3$; M = C,Si,Ge,Sn,Pb)

k_{rel}	C	Si	Ge	Sn	Pb
	1	10^4	10^6	10^{10}	10^{13}

[79,80] (nucleophile vs. electrophile) can be attributed to hyperconjugation.

In the above survey, which of course is not exhaustive, a series of examples was given in an attempt to interpret the observed physical or chemical phenomena in terms of hyperconjugation. Many of the arguments presented here, however, are based only on rough qualitative ideas, and frequently the relationship of the observed experimental quantities to hyperconjugation is only intuitive. In order to prevent possible misunderstandings and misinterpretations concerning hyperconjugation, a thorough analysis of the theoretical aspects of the resonance phenomenon is included in a subsequent section.

4.3.3 Theoretical Aspects of Hyperconjugation

From the theoretical point of view the only convincing proof of resonance interactions consists of the appearance of deviations from strict adherence to molecular additivity rules. The existence of such interactions is conditioned by the possibility of describing molecular electronic structure in terms of certain local contributions corresponding to individual chemical bonds. Since the quantum chemical description of molecular structure does not require the introduction of the concept of the chemical bond, it is necessary to note the mutual relationship of the classical and the quantum model of molecular structure.

Instead of relatively independent chemical bonds localized in different regions of space, quantum chemistry operates with the concept of a wave function spread over the whole molecule. In the early days of quantum chemistry, attempts were made to show a mutual relationship between both types of descriptions. As a result of this effort, it was confirmed that despite apparent differences there is a close parallelism between the quantum and the classical pictures of chemical bonding. Such parallelism originates in the fact that the total SCF wave function is invariant with respect to arbitrary unitary transformation between the set of occupied molecular orbitals [81,82]. The form of the transformation matrix can sometimes be determined from the symmetry of the studied problem. Generally, however, it is necessary to use some external

THEORETICAL ASPECTS OF BONDING

physical criterion. An exhaustive review of different localization methods can be found in the literature [83]. The actual calculations performed by different localizations show that the form of resulting localized orbitals is to a considerable extent insensitive to the chosen localization criterion. The localized orbitals obtained in this way are entirely equivalent to the original canonical molecular orbitals because they do not alter the total wave function of the molecule. However, since each of them is localized in a region of a certain chemical bond in a molecule, they allow a chemically more illustrative interpretation of molecular structure in terms of different local contributions.

As demonstrated by Dewar [84], these localized orbitals can be used only for calculation of the "collective" molecular properties such as charge distribution, heats of formation, dipole moments, etc. In addition to "collective" properties, Dewar also distinguishes "one-electron" properties, such as ionization potential, UV excitation energy, odd electron distribution in radicals, etc., the calculation of which requires the use of the original delocalized canonical molecular orbitals. In this case, additive schemes are not applicable even in a very crude approximation, as a consequence of the fact that, e.g., the ionization or excitation of the molecule cannot be considered as a local process occurring specifically on some bonds but rather on the molecule as a whole. The localized orbitals determined according to some of the usual criteria are entirely equivalent to the original canonical MO in describing the "collective" molecular properties, but, strictly speaking, they cannot be considered as an exact counterpart of the classical concept of the chemical bond. This distinction occurs because each of these orbitals is not localized entirely in the region of a certain bond in a molecule; it always contains (very important!) some small "tails" over the remaining part of the molecule.

The classical concept of the chemical bond has its exact theoretical counterpart in strictly localized orbitals. These orbitals represent an artificial bonding model constructed under the restrictive condition that they are localized only between atoms connected by a formal chemical bond. The wave function constructed from these strictly localized orbitals then exactly corresponds to the structure characterized by the classical structural

formula. Such a strictly localized wave function does not
describe exactly the real molecular structure, since this
would imply an exact validity of molecular additivity
rules. Moreover, it would mean that the real molecular
structure can be expressed by only one classical structural formula. The impossibility of such a description
was recognized for a special class of aromatic systems by
classical valence theory, and it led to the concept of
resonance. It should be realized, however, that resonance
does not represent any real physical phenomenon; it
appears only as a consequence of the imperfect description
of molecular structure in terms of localized chemical
bonds. The concept of resonance could be entirely avoided
when describing molecular structure in terms of a wave
function. Such a description, however, can never reach the
elegance and simplicity of classical structural formulae,
which definitely remain the basic tool of representation
of chemical structure even at the expense of preserving
such auxiliary concepts as resonance and hyperconjugation.

From the above discussion it follows that, like
$(p-d)_\pi$ interaction, the hyperconjugation concept is also
to be considered only a means of description and not an
interpretation of the observed phenomena. It is certainly
true that valence theory can contribute to the elucidation
of many misunderstandings concerning the concept of hyperconjugation simply by determining the range of validity
and applicability of a localized model of chemical structure. A convincing experimental proof of resonance
interactions in a series of saturated molecules is hampered by the small extent of deviations from perfect additivity of molecular properties. For that reason it is
better to approach the problem of the determination of the
extent of hyperconjugation theoretically by comparing the
"exact" wave function of the molecule with the hypothetical strictly localized wave function corresponding to the
classical structural formula. The "similarity" of these
two functions then characterizes, in certain respects, the
extent of resonance interactions. Such an approach is consistent with the recently proposed criterion for the
objective assessment of the extent of hyperconjugation in
organosilicon chemistry [85].

4.3.4 Hyperconjugation and the Accuracy of Localized Descriptions of Bonding

Let the molecular structure of a given molecule be described by the set of canonical molecular orbitals Φ_i. It is of no importance whether these orbitals were obtained on the ab initio level or with the aid of some semi-empirical method. The ground state SCF wave function is then given by the Slater determinant ϕ_o (Eq. 20).

$$\phi_o = |\Phi_1\bar{\Phi}_1\phi_2\bar{\phi}_2\cdots\Phi_N\bar{\phi}_N| \qquad (20)$$

Suppose that this wave function describes the real electronic structure of a molecule. In order to characterize the extent of delocalization interactions in a molecule, it is first necessary to define some standard reference structure in which the delocalization does not take place. The wave function of this standard is given by the Slater determinant Λ_o (Eq. 21) constructed from strictly localized orbitals λ corresponding to individual chemical bonds.

$$\Lambda_o = |\lambda_1\bar{\lambda}_1\lambda_2\bar{\lambda}_2\cdots\lambda_N\bar{\lambda}_N| \qquad (21)$$

The basic idea of the construction of this localized wave function thus resembles the so-called PCILO method [86]. There is, however, one important difference in that the concrete form of strictly localized orbitals is determined as in loge theory [87] from the condition of the minimal electronic fluctuation in each of the bonds. Such a condition was mathematically formulated by Polák [88,89] in the form of the maximalization of orbital occupation numbers n_μ in occupied localized orbitals λ_μ. The physical content of this condition can be expressed simply by the following considerations.

The localized wave function Λ_o is not the eigenfunction of the molecular Hamiltonian. From this it follows that in a state described by the function Λ_o the energy of the system has no sharp value; it fluctuates. The physical meaning has only the mean value of energy given by Eq. (22).

$$\bar{E} = \langle\Lambda_o|\hat{H}|\Lambda_o\rangle \qquad (22)$$

A similar situation also exists for other observables. For example, the number of electrons in a state described by the function Λ_o is also not constant but fluctuates with time. The physical meaning corresponds only to the mean number of electrons in a given structure. Since the hypothetical structure Λ_o is composed of independent localized bonds, it is evident that fluctuation also exists in the number of electrons per bond. As a consequence of this fact, it is no longer possible to place exactly two electrons in a bond. Such a conclusion is entirely analogous to the conclusions of loge theory about the fluctuation of electrons in loges. The mean number of electrons in a bond μ (in orbital λ_μ) can be calculated from the well-known formula (Eq. 23),

$$n_\mu = 2 \sum_i^{occ} L_{\mu i}^2 \qquad (23)$$

provided that one knows the expansion coefficients L of molecular orbitals ϕ in the basis of strictly localized orbitals λ. The matrix of expansion coefficients L can be expressed from the basis of LCAO expansion coefficients C (Eq. 24).

$$\psi = \chi C \qquad (24a)$$

$$\lambda = \chi A \qquad (24b)$$

$$\Phi = \lambda L = \chi AL \qquad (24c)$$

$$L = A^{-1} C \qquad (24d)$$

As can be seen from the last equation, the final matrix L depends on the values of the expansion coefficients A, which determine the form of the strictly localized orbitals λ. Polák [88,89] has proposed a method of variational optimization of matrix A, thus ensuring the maximalization of occupation numbers n_μ.

The values of occupation numbers can be roughly divided into two groups. In the first, corresponding to bonding orbitals λ_μ, the values of n_μ are close to two. In contrast, the values of n_μ^* in antibonding orbitals are close to zero. The mean number of electrons in a given

THEORETICAL ASPECTS OF BONDING 259

structure is given by the simple sum of occupation numbers n_μ in localized bonding orbitals λ. It is evident that this sum is always smaller, or at a maximum, equal to the total number N of electrons in a molecule. The accuracy of localized description can be characterized, in the framework of the present approach, by expression (25)

$$\xi = \frac{\sum_\mu^{occ} n_\mu}{N} \quad (25)$$

The value of coefficient ξ varies between 0 and 1, and the closer is its value to unity, the better is the molecular structure depicted by the localized wave function Λ_o. Actual values of parameter ξ calculated in the framework of a CNDO approximation for a series of several silicon and analogous carbon compounds are collected in Table II. As can be expected, the ordinary structural formula (corresponding to the localized function Λ_o) describes the molecular structure of saturated or nonconjugated molecules with remarkable accuracy. Nevertheless, there are some general features worth mentioning.

The classical chemical picture of hyperconjugation suggests that the substitution in a molecule by a methyl or a silyl group leads to a decrease of accuracy of the localized model, and this decrease is generally larger in a silyl than in a methyl series. The same conclusion follows from the calculations summarized in Table II. Moreover, there is a much more pronounced decrease of accuracy of the localized model in a series of n-polysilanes than in the analogous n-alkanes. This result is consistent with the recently described "aromatic" behavior of polysilanes [89a].

4.4 ELECTRONEGATIVITY

One of the theoretical concepts frequently used to interpret the structure and reactivity of organosilicon compounds is the concept of electronegativity. The larger +I effect of the silyl substituents, which can be seen in the acidity of a series of carboxylic acids RCH_2COOH [90] (R = H, CH_3, CMe_3, $SiMe_3$), was interpreted by such argu-

Table II. Calculated values of parameter ξ (Eq. 25) characterizing the accuracy of the localized model of chemical structure in a series of some analogous carbon and silicon compounds

Compound	ξ	Compound	ξ
CH_4	0.999	SiH_4	0.996
C_2H_6	0.995	Si_2H_6	0.992
C_3H_8	0.994	Si_3H_8	0.980
$CH_2=CH_2$	0.994		
$CH_3CH=CH_2$	0.992	$SiH_3CH=CH_2$	0.991
$CH\equiv CH$	0.998		
$CH_3C\equiv CH$	0.992	$SiH_3C\equiv CH$	0.993
H_2O	0.998		
CH_3OH	0.995	SiH_3OH	0.991
CH_3CH_2OH	0.993	SiH_3CH_2OH	0.993
NH_3	0.999		
CH_3NH_2	0.995	SiH_3NH_2	0.993
$CH_3CH_2NH_2$	0.993	$SiH_3CH_2NH_2$	0.992
HF	0.999		
CH_3F	0.996	SiH_3F	0.992
CH_3CH_2F	0.994	SiH_3CH_2F	0.994

ments. The lower electronegativity of silicon with
respect to carbon is also considered to be the reason for
the easy solvolysis of SiH bonds. In all of these
interpretations the electronegativity was generally considered in the sense of the classical Pauling definition
as an invariant property of the element, independent of
whether the corresponding atom is isolated or bound in a
molecule. An increasing number of results, however, indicate that such an interpretation of electronegativity
is not consistent with experimental results. The values of
electronegativity, for example, vary with the oxidation
state of the atom in a molecule. In light of these findings it is not surprising that some experimental results
of the reactivity and spectral behavior of carbon-
functional organosilicon compounds contradict the original
Pauling values of the respective electronegativities of
carbon and silicon. For example, the ^{35}Cl nuclear quadru-
pole resonance (NQR) spectra of (chloromethyl) substituted
silanes indicate that silicon exhibits, in comparison with
carbon, a strongly electron-accepting behavior [91]. Similarly, the electron-accepting behavior of silyl groups
also appears in the solvolysis of (chloromethyl) silanes
[92]. It is evident that when attempting an interpretation of these results in terms of classical Pauling
electronegativity, difficulties arise that can be overcome
only by admitting the anomalous behavior of silyl groups.
A discussion of this anomalous behavior necessitates the
introduction of the term "the α-effect".

By using a more "convenient" definition of electronegativity, one that respects the dependence of electronegativity on the actual valence state of the atom in a
molecule, it was shown that it is possible to explain the
electron-accepting behavior of silyl substituents [93].
This view of electronegativity originates from the earlier
Hinze and Jaffé idea of electronegativity as an orbital
property [94] and now generalized to the level of a
semiempirical CNDO approximation [93]. This generalization
is based on the possibility of the partitioning of total
molecular energy into mono- and biatomic components (Eq.
26).

$$E = \sum_A \varepsilon_A + \sum\sum_{A<B} \varepsilon_{AB} \qquad (26)$$

The original expression, devised by Pople and coworkers [95], was modified in such a way that the resulting monoatomic terms depend only on the charge densities $p_{\mu\mu}$. The detailed procedure of the reorganization of the original partitioning formula can be found in the original paper [93]. In this way it was possible to obtain the expressions that properly respect the dependence of orbital electronegativites on the structure of the molecule (Eq. 27).

$$\chi_j^A = -\frac{\partial \varepsilon_A}{\partial p_{jj}} = -U_{jj}^A - (P_A - \frac{1}{2})\gamma_{AA} \qquad (27)$$

On the basis of orbital electronegativities it is possible to define the global atomic electronegativity of the atom in a molecule (Eq. 28).

$$\delta = \frac{\sum_j^A P_{jj}\chi_j^A}{P_A} \qquad (28)$$

This definition leads to isolated atom electronegativites that are essentially equivalent to the original Pauling values. The main advantage of the new definition is in allowing a proper description of the changes in electronegativity caused by the change in the atomic valence state on going from one molecule to another. To illustrate these conclusions the electronegativity values δ for a series of several silicon and analogous carbon compounds are presented in Table III. As can be seen from this Table, the electronegativity of the isolated silicon atom in the ground state, as well as in the sp^3 valence state, is lower than the electronegativity of carbon. On the other hand, the order of electronegativities of both elements is reversed in silane and methylsilane. The same order is also found in a comparison of α-carbon-functional trimethylsilylmethyl derivatives with analogous ethyl derivatives. The increased electronegativity of silicon in α-carbon-functional derivatives thus explains the observed electron-accepting character of silyl substituents in these compounds. These conclusions demonstrate

Table III. Calculated values of atomic electronegativity δ of carbon and silicon in some simple molecules

Compound	δ^C(eV)	Compound	δ^{Si}(eV)
C (s^2p^2)	9.81	Si (s^2p^2)	7.08
C (sp^3)	7.69	Si (sp^3)	5.60
CH_4	7.00	SiH_4	11.28
CH_3SiH_3	5.10	SiH_3CH_3	11.43
*CH_3CH_2F	6.98	Me_3SiCH_2F	11.04
*CH_3CH_2OH	7.31	Me_3SiCH_2OH	10.95
*$CH_3CH_2NH_2$	7.43	$Me_3SiCH_2NH_2$	11.04

that the interpretation of experimental results in terms of electronegativity requires a more extensive theoretical analysis, since the original classical ideas do not necessarily describe the physical nature of the observed phenomena properly.

4.5 THE ELECTRONIC EFFECTS OF SILYL SUBSTITUENTS AND THE POSSIBILITIES OF THEIR CHARACTERIZATION

4.5.1 Linear Free Energy Relationships

The Hammett and Taft equations, which are well known representatives of LFER, are widely used to describe substituent effects in classical organic chemistry and also in organosilicon chemistry in a semiquantitative manner. Two types of problems are encountered in applying the above two equations in organosilicon chemistry. The first concerns the application of LFER to systems containing silicon as a reacting center. Reactions belonging to this category can be roughly divided into two groups according to whether or not a splitting of the silicon-carbon bond takes place. The first class of reactions is represented by protodesilylation reactions where the values of reaction constant ρ were interpreted in terms of $(p-d)_\pi$ interaction [69]. An alternative explanation was proposed in the preceding section. Among reactions proceeding without scission of the Si-C bond, the most frequently studied were the solvolysis of silanes [96-98] and the insertion of carbenes into the Si-H bond [99] (Eq. 29).

$$X-C_6H_4-\underset{\underset{Me}{|}}{\overset{\overset{Me}{|}}{Si}}-H \xrightarrow{PhHgCCl_2Br} X-C_6H_4-\underset{\underset{Me}{|}}{\overset{\overset{Me}{|}}{Si}}-CCl_2H \qquad (29)$$

The second problem concerning the application of LFER in organosilicon chemistry is connected with the characterization of the substituent effect of the silyl substituents.

On the basis of early ideas about the lower electronegativity of silicon with respect to carbon, one could expect the silyl substituents to exhibit more extensive electron-donating properties in comparison with analogous carbon groups. Numerous experimental results are consistent with such a simple picture. For example, Sommer [90] explains the decreasing order of acidity of carboxylic acids XCH_2COOH (Eq. 30) in such a way.

$$X \equiv H > CH_3 > (CH_3)_3C > (CH_3)_3Si \qquad (30)$$

The same order of acidity also appears in a series of substituted toluylic acids $XCH_2C_6H_4COOH$ [100]. The simple concept of the successive attenuation of the silicon +I effect with an increasing distance of the silicon atom from the reacting center is also consistent with the order of basicity of trimethylsilyl substituted alcohols, $Me_3Si(CH_2)_nOH$ [101].

These simple ideas have to be significantly modified, however, in cases where the silyl groups are directly attached to an aromatic nucleus, in which case the substituent effect is characterized by the value of the so-called σ constant. A number of authors have attempted to determine these constants for silyl groups [102-108]. The

THEORETICAL ASPECTS OF BONDING 265

results of their efforts demonstrate that the electron donating effect of these groups is partially or sometimes completely compensated by the mechanism of the back-donation (the -M effect) that was frequently attributed to $(p-d)_\pi$ interaction. The relative extent of +I and -M effects of silyl substituents (or, in other words, the resulting "net" substituent effect of these groups) is not, however, constant; it is influenced by the electronic requirements of the particular reaction. Thus, for example, the negative sign of the σ_p constant of the trimethylsilyl group, as determined from the dissociaton of substituted benzoic acids [103], implies that the +I effect of silicon is only partially compensated by back-bonding while the σ_p value for the same Me_3Si group resulting from basicities of substituted anilines [104] suggests a strong electron accepting behavior for this group. The dual character of silyl substituents was similarly demonstrated in an aliphatic series [79].

On the basis of these results it is evident that the appearance of electron-accepting behavior of silyl groups is not restricted to the structural arrangement RR'R"SiX (X = hal, OR, NR_2, Ph) but it appears also in cases where the functional group X is bonded to silicon through one or more intervening CH_2 groups. The confirmation of this conclusion arises from basicity measurements in a series of trimethylsilyl substituted amines $Me_3Si(CH_2)_nNH_2$ [109] and ethers $H_3M(CH_2)_nOCH_3$ [110,111] (M = C, Si). Similarly, as indicated by NQR spectrometry, a decrease of electron density on the chlorine atom in (chloromethyl) substituted silanes [91,112] is observed. Such anomalies, when n = 1, are termed "the α-effect". The mechanism of this effect is still the subject of discussion, and several alternatives have been proposed for its explanation [112-114]. The only conclusion about the mechanism of the α-effect that seems to be certain is that its existence does not require the bonding utilization of silicon d orbitals [115]. A more detailed discussion of the mechanism of the α-effect will be given later.

In addition to the electron-accepting behavior of silyl groups that is characteristic for the α-effect, one also observes in certain cases a strong electron donating effect, as for example in benzylsilanes [46,102,116,117] or allylsilanes [44]. The anomalous reactivity of these compounds has given rise to the term "β-effect" [118].

All these examples clearly show that the substituent effect of silyl groups cannot be unequivocally characterized, since the resulting behavior of polarizable silyl substituents is influenced considerably by numerous external factors, e.g., by molecular conformation [119] or by varying electronic demands in the transition states of different reactions [78].

From this it follows that for an adequate description of the electronic effects of these substituents it is necessary to abandon the global characterization that forms the basis of the LFER approach and to analyze the detailed factors contributing at a microscopic level to the final "total" substituent effect. The most convenient means of such an analysis would be, of course, the quantum chemical theory of substituent effects [120], but up to now there has been no attempt to apply it to organosilicon chemistry. Besides, several simpler methods have been recently proposed and applied to the analysis of electronic substituent effects. One of them is represented by a comparison of the stability of substituted molecules in some properly chosen isodesmic reactions. This method, which is more convenient than a direct comparison of molecular energies because it partially compensates for the differences in correlation energy, was used to investigate the relative ability of methyl and silyl substituents to stabilize the adjacent anionic center [121]. The proper isodesmic process is characterized in this case by Eq. (31).

$$XCH_2^- + CH_4 \longrightarrow XCH_3 + CH_3^- \qquad (31)$$

$$X = CH_3, SiH_3$$

The calculations imply that the silyl group is significantly more effective in stabilizing the α-anionic center than is the methyl group. This conclusion seems to demonstrate nearly unequivocally the operation of a $(p-d)_\pi$ interaction. Nevertheless, the calculations clearly document that such is not the case, since the same conclusions are obtained even when neglecting silicon d orbitals [121].

Another possibility of studying substituent effects is represented by Pople's method of analysis of the curves

THEORETICAL ASPECTS OF BONDING

of internal rotation [122,123]. Its detailed application to carbon-functional organosilicon compounds is the subject of the subsequent section.

4.5.2 Fourier Component Analysis of Internal Rotation

The form of the potential curves of internal rotation is determined by the extent and the nature of intramolecular interactions in a given molecule. The knowledge of the rotational potential thus enables one to analyze these interactions. One of the possibilities of such analysis is represented by the method of a Fourier component analysis of internal rotation introduced by Pople and coworkers [122,123]. The potential of internal rotation is expressed in this method in the form of a truncated Fourier series where the coefficients of the expansion are given a certain physical meaning. In characterizing the substituent effect of the substituent in $Y-CH_2-X$ derivatives the concrete form of Fourier expansion is described by Eq. (32).

$$V(\phi) = \frac{V_1}{2}(1 - \cos\phi) + \frac{V_2}{2}(1 - \cos 2\phi) + \frac{V_3}{2}(1 - \cos 3\phi) \quad (32)$$

In this equation the coefficient V_1 characterizes the extent of dipole-dipole interaction between the dipole of the C-Y bond and the dipole of free electron pairs on substituent X. In addition to the extent of the interaction, the sign of coefficient V_1 also enables one to deduce the relative orientation of both dipoles. Coefficient V_2 describes the extent of delocalization interactions corresponding to the electron transfer from the free pair at X to the antibonding orbital of the C-Y bond. In a certain sense this contribution represents hyperconjugation. Finally, coefficient V_3 represents, in a series of $Y-CH_2-X$ compounds with constant X, a measure of the steric effect of substituent Y. On the other hand, for a constant Y substituent in the same series, its magnitude corresponds to the number of H...H nonbonding interactions. Thus, in a series $X-CH_3$, NH_2, and OH, its magnitude decreases roughly in a 3:2:1 ratio regardless of the type of Y substituent.

This method of analysis of intramolecular interactions

was applied to the detailed elucidation of the substituent effect of silyl substituents in α and β carbon-functional derivatives [124,125]. The potential of internal rotation was calculated in the so-called "rigid rotor approximation" by the standard CNDO/2 method both including and neglecting silicon d orbitals. The calculated curves for rotation around the C-N bond in ethylamine and silylmethylamine are presented for illustration in Fig. 3. The chemical similarity of carbon and silicon manifests itself in the similarity of the corresponding potential curves. This similarity, however, appears only when d orbitals are neglected at silicon. On the other hand, the inclusion of d orbitals leads to an entirely unrealistic rotational potential curve. For that reason the interpretation of the substituent effect of silyl substituents was done only with calculations neglecting d

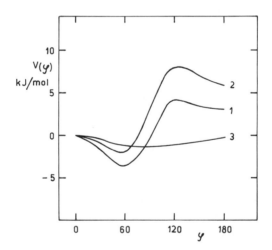

Fig. 3. CNDO/2 calculated potentials of internal rotation around the C-N bond in α-functional amines:

1 $CH_3CH_2NH_2$,

2 $H_3SiCH_2NH_2$ without d orbitals,

3 $H_3SiCH_2NH_2$ with d orbitals.

orbitals. Calculated values of coefficients V_1-V_3 for α-functional organosilicon alcohols and amines are collected in Table IV.

Attention should now be focused on a more detailed discussion of coefficients V_n. As follows from the sign of coefficient V_1, the dipole of the C-Y bond is polarized in the direction Y→C for all substituents except Y = F. Comparison of a series Y = CH_3, SiH_3, and $SiMe_3$ shows that the magnitude of this dipole increases in the order CH_3 < SiH_3 < $SiMe_3$, which is consistent with the expected order of the +I effect of these substituents. This donating effect is partially compensated by the backdonation from the lone pairs of OH or NH_2 groups, respectively, into the σ^*_{C-Y} antibonding orbital of the C-Y bond. The extent of this backdonation is determined by coef-

Table IV. Calculated values of Fourier coefficients V_n (kJ/mol) for internal rotation around the CX bond in α-carbon-functional derivatives Y-CH_2-X

Compound	V_1	V_2	V_3
CH_3NH_2	-	-	-6.56
FCH_2NH_2	-3.22	3.59	-5.77
$CH_3CH_2NH_2$	9.69	-3.40	-6.51
$SiH_3CH_2NH_2$	16.61	0.29	-8.41
$SiMe_3CH_2NH_2$	17.47	0.46	-8.89
CH_3OH	-	-	-3.28
FCH_2OH	7.21	-2.86	-2.74
CH_3CH_2OH	-8.50	2.34	-3.06
SiH_3CH_2OH	-20.67	-0.93	-4.33
Me_3SiCH_2OH	-20.94	-0.79	-4.54

ficient V_2. Calculated values suggest that this backdonation decreases in the series $F > SiH_3 > SiMe_3$. From the sign of coefficient V_2 for $Y = CH_3$ it follows that in this case backdonation does not take place. On the contrary, the electronic effects parallel the +I effect and can be characterized as $\sigma_{C-CH_3} \to \sigma^*_{NH}$ or $\sigma_{C-CH_3} \to \sigma^*_{OH}$. The negative sign of coefficient V_3 implies in all cases a preference of staggered over eclipsed conformation. Its magnitude characterizes the steric effect of Y substituents in the individual series of alcohols and amines.

The intramolecular interactions in β-carbon-functional derivatives were analyzed in a similar way. In studying the rotation around the C-X bond these compounds can be considered as α-functional. The only difference consists in the MH_3CH_2 substituent exhibiting a more complex internal structure, which manifests itself in the existence of two energetically comparable conformational arrangements that are depicted in Fig. 4. A Fourier component analysis of the curve of internal rotation leads in the case of these β-carbon-functional derivatives to the conclusion that the electronic effect of silyl substituents depends, in contrast to the effect of simple alkyls, on the molecular conformation. Thus, the effect of a SiH_3 group placed into the "anti" position to substituent X (Fig. 4a) is practically the same as the effect of a CH_3 group. On the other hand, placing the same substituent in the "gauche" position (Fig. 4b) leads to a considerable increase of the electron donor

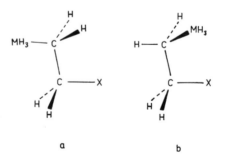

Fig. 4. Two energetically comparable conformational arrangements of $H_3M(CH_2)_2X$ molecules.

THEORETICAL ASPECTS OF BONDING 271

effect. In contrast, for analogous carbon derivatives the effect is practically the same as in the "anti" position. These results document that the substituent effect of silyl substituents is crucially influenced by the molecular conformation. This conformational dichotomy of the substituent effect, which indicates increased polarizability of silicon, has important consequences for the mechanism of the α-effect.

4.5.3 The α-Effect and Acid-Base Properties[x]

Numerous experimental data suggest that despite the expected close chemical similarity of carbon and silicon the electronic effects of silyl groups are substantially different from those of simple alkyls. Thus, in a series of trimethylsilyl substituted alcohols, $Me_3Si(CH_2)_nOH$, one observes a regular decrease of basicity with increasing length of the alkyl chain [101]. On the other hand, an entirely different order of basicity is observed for the analogous amine series where the maximum basicity occurs for the β-functional derivative [109]. This order of basicity was rationalized in terms of the α-effect. Because of unavailability of direct gas phase techniques [126,127], the basicity in these studies was measured by the infrared method with phenol or deuterochloroform as proton donors. Since the measurements were made in a very dilute solution in nonpolar CCl_4, and since the studied series form a structurally related set of compounds, the IR data can be regarded as a rough measure of gas phase basicities.

The theoretical interpretation of the observed trends in basicities in both series of compounds is based on the calculation of the protonation energies. They were calculated for all $H_3M(CH_2)_nX$ derivatives (M = C, Si; X = NH_2, OH; n = 1,2,3) [119] in their most stable conformations, which are depicted in Figs. 4 and 5. Results of these calculations are summarized in Table V. As follows from the values of the protonation energies in a series of aliphatic alcohols and amines, the CNDO/2 calculations correctly reproduce the experimentally found increase in

---

[x] A more complete review of experimental results related to the α-effect can be found in Chapter 2.

anti-anti   gauche-gauche

Fig. 5. Anti-Anti and Gauche-Gauche Comformations of $H_3M(CH_2)_2X$ Molecules.

basicity with the increasing length of the alkyl chain, irrespective of conformation. On the other hand, the effect of the SiH_3 group in organosilicon derivatives is dramatically different. This difference is demonstrated for representative series of alkyl and silylalkyl alcohols in Fig. 6. Essentially the same picture occurs with the analogous amines. As shown by Fig. 6, the trend in protonation energies in a series of silyl derivatives such as SiH_3CH_2OH, gauche $SiH_3CH_2CH_2OH$, and gauche-gauche $SiH_3CH_2CH_2CH_2OH$, corresponds to a regular decrease of basicity. In contrast, in a similar series of trans derivatives, the trend of basicity is much more complex and in fact reproduces the experimental results characteristic of the α-effect.

Theoretical calculations thus suggest that the appearance of the α-effect is crucially connected with the anti conformation of the molecular chain. In this connection it would certainly be interesting to verify such theoretical conclusions by the experimental study of gas phase basicities on properly chosen rigid models represented, e.g., by compounds E and F.

THEORETICAL ASPECTS OF BONDING

Table V. CNDO/2 calculated protonation energies (kJ/mol) in a series of $H_3M(CH_2)_nX$ carbon-functional derivatives

Compound	ΔE	
	M = C	M = Si
$H_3MCH_2NH_2$	-1288.7	-1305.0
$H_3MCH_2CH_2NH_2$[a]	-1293.6	-1292.6
$H_3MCH_2CH_2NH_2$[b]	-1299.6	-1318.7
$H_3MCH_2CH_2CH_2NH_2$[c]	-1296.9	-1298.0
$H_3MCH_2CH_2CH_2NH_2$[d]	-1304.9	-1311.9
H_3MCH_2OH	-1103.5	-1123.6
$H_3MCH_2CH_2OH$[a]	-1112.4	-1118.1
$H_3MCH_2CH_2OH$[b]	-1117.0	-1141.7
$H_3MCH_2CH_2CH_2OH$[c]	-1116.0	-1117.0
$H_3MCH_2CH_2CH_2OH$[d]	-1123.4	-1132.0

[a]conformation gauche, [b]conformation anti, [c]conformation gauche-gauche, [d]conformation anti-anti.

The origin of this stereospecificity of the electronic effects of silyl substituents was recently analyzed on the basis of the decomposition of the total protonation energies into components corresponding to different mechanisms of intramolecular interaction [128]. The detailed method of decomposition is described in the original paper [129]. The protonation energy ΔE is decomposed into three additive components (Eq. 33).

$$\Delta E = \Delta \epsilon + \Delta E_{elst} + \Delta E_{polar} \quad (33)$$

The first term, $\Delta\varepsilon$, represents a correction of the extent of the basis set analogous to the function counterpoise correction [130]. The second and the third terms correspond to electrostatic and polarization contributions. The results of the decomposition of a series of silyl substituted carbon-functional alcohols and amines are summarized in Table VI. A comparison of individual components in a series of derivatives such as SiH_3CH_2X and "anti" $SiH_3CH_2CH_2X$ clearly documents that the increase in basicity characteristic of this conformation of the chain is primarily due to an decrease of the electrostatic contribution. On the other hand, in a comparison of SiH_3CH_2X and "gauche" $SiH_3CH_2CH_2X$, the increase in electrostatic contribution is responsible for the decrease of basicity.

These results thus clearly suggest that the α-effect is probably of electrostatic origin. This may explain the fact that its appearance is not restricted only to basicity but that it is also displayed by molecules in their ground states. (See, e.g., NQR evidence in [91,131]).

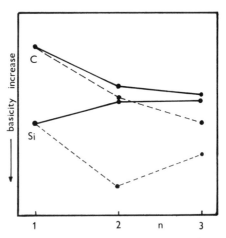

Fig. 6. Dependence of calculated protonation energies in a series of $H_3M(CH_2)_nOH$ carbon-functional derivatives on the value of n; solid line = gauche conformations; broken line = anti conformations.

Table VI. Protonation energies and their individual components in a series of α- and β-carbon-functional alcohols and amines (all quantities in a.u.)

Compound	ΔE	$\Delta \varepsilon$	ΔE_{elst}	ΔE_{polar}
$SiH_3CH_2NH_2$	-0.497	-0.385	0.049	-0.161
$SiH_3CH_2CH_2NH_2$[a]	-0.502	-0.383	0.042	-0.161
$SiH_3CH_2CH_2NH_2$[b]	-0.492	-0.388	0.058	-0.162
SiH_3CH_2OH	-0.428	-0.387	0.100	-0.141
$SiH_3CH_2CH_2OH$[a]	-0.435	-0.385	0.093	-0.143
$SiH_3CH_2CH_2OH$[b]	-0.424	-0.391	0.110	-0.143

[a] anti conformation

[b] gauche conformation

4.6 THE CHEMISTRY OF SILICENIUM IONS AND SILYL ANIONS

4.6.1 Silicenium Ions

The existence of silicenium ions as intermediates has been frequently proposed by various authors for numerous reactions in organosilicon chemistry [132]. All experimental attempts to detect their existence in solution [133,134] have failed, however, despite the fact that under comparable conditions the existence of the analogous carbonium ions has been convincingly proved [135,136]. The results of a variety of gas phase studies, on the other hand, clearly suggest that R_3Si^+ ions appear frequently as the product of fragmentation in mass spectra of different organosilicon compounds [137]. Silicenium ions were also detected in the gas phase under the conditions of ion cyclotron resonance, which, of course, implies that the unsuccessful attempts to detect the existence of silicenium ions in solution are not due to an inherent instability of these species.

The stability of silicenium ions was recently studied by ab initio quantum chemical methods. In addition to a comparison of the stability of the parent CH_3^+ and SiH_3^+ ions, the effect of substituents was also investigated. Thus, for example, Apeloig and Schleyer [138] studied the relative stability of SiH_2X^+ and CH_2X^+ ions in the STO-3G basis. The difference in stability of these ions is characterized by the energy of the isodesmic reaction (Eq. 34).

$$SiH_2X^+ + CH_3X \dashrightarrow CH_2X^+ + SiH_3X \qquad (34)$$

Its maximal value (305 kJ/mol in favor of silicenium ions) is reached for the parent SiH_3^+ and CH_3^+ ions. All other substituents lead to a decrease of this value, and this decrease is especially large for π donors; thus, for example, $NH_2SiH_2^+$ is more stable than $NH_2CH_2^+$ by only 75 kJ/mol. The higher stability of silicenium ions in comparison with carbonium ions is not surprising since the silicon atom, due to its greater polarizability, is more capable of stabilizing the positive charge than is carbon. The substituent effect of the X substituent on the stability of MH_2X^+ ions with respect to parent MH_3^+ species can be characterized by means of the isodesmic reaction described by Eq. (35).

$$MH_2X^+ + MH_4 \dashrightarrow MH_3^+ + MH_3X \qquad (35)$$

The calculations suggest that all substituents stabilize the cationic center more effectively than hydrogen and that this stabilization is relatively higher in the carbon than in the silicon series, which is again a consequence of the greater polarizability of silicon.

The theoretical conclusions in the work of Apeloig and Schleyer are in complete agreement with available experimental data on the stability of the silicenium ions obtained by photoionization mass spectrometry [139]. Thus, the instability of silicenium ions in solution can be attributed to the difference in the heterolytic dissociation energy of the M-X bond in R_3M-X (M = C, Si; X = F and probably also OH and Cl) derivatives. The dissociation energy is higher for silicon derivatives by approximately 210 kJ/mol.

4.6.2 Silyl Anions

The relative abilities of carbon and silicon to stabilize a negative charge have recently been the subject of both experimental [140-142] and theoretical studies [121]. As demonstrated by ab initio calculations, the proton affinity of the silyl anion SiH_3^- is smaller than that of the methyl anion. Their difference equals roughly 270 kJ/mol on the level of the double zeta basis and decreases to 210 kJ/mol with the inclusion of polarization functions [121]. This clearly demonstrates that the silicon, again due to its greater polarizability, can better accomodate the negative charge than can carbon. The greater polarizability of silicon also means that whereas the stabilization of the negative charge in the triphenylmethyl anion proceeds through conjugation with phenyl rings, the conjugation in the corresponding silicon analogue is negligible [141,142]. As demonstrated by the results of ab initio calculations, the introduction of a methyl substituent into a silyl anion even leads to a decrease of its proton affinity, which indicates that the methyl substituent destabilizes the silyl anion [121].

The relative ability of CH_3 and SiH_3 substituents to stabilize the adjacent anionic center can be compared on the basis of the isodesmic reaction given in Eq. (31). The calculations prove that the silyl group is more effective in stabilizing the anionic center than is the methyl group. This greater stabilization cannot, however, be considered as a proof of d orbital participation since the same conclusions hold even if d orbitals are neglected at silicon. A more convenient description of this result thus seems to be hyperconjugation [121].

4.7. COMMON ASPECTS OF CHEMICAL REACTIVITY OF CARBON-FUNCTIONAL ORGANOSILICON COMPOUNDS

The α- and β- derivatives occupy an important place among carbon-functional organosilicon compounds. Their chemical reactivity is in some cases so remarkable that it has given rise to the term α- and β-effect [45,92,118]. It appears, however, that when classifying the chemical properties of carbon-functional derivatives, it is preferable not to distinguish α- and β-effects in the sense that the α-effect is observed in α-functional and β-effect in β-functional compounds; this straightforward classification may hide the common features of both effects. It is much better to focus on the following common features.

1) Carbon-functional derivatives: X_3Si-CH_2-Y, $X_3Si-CH=Y$ (Y = hal, OR, NR_2, SR, =O, etc.)

Due to the presence of the electronegative substituent, the α carbon bears a partial positive charge. At the same time the free electron pair appears in the β position to silicon.

2) Carbon-functional derivatives: X_3Si-CH_2-Y, $X_3Si-CH_2-CH_2-Y$ (Y = COR, COOR, CN, hal, OR, NR_2)

Due to the electronegative substituent, the positive charge appears in the β position to silicon. The consequences of non-uniform charge distribution on chemical reactivity can be summarized as follows.

a) The Presence of a Positive Charge on the α Carbon

Since the silyl group is less effective in stabilizing the adjacent cationic center than an alkyl group, migration from silicon to the positively charged carbon frequently takes place. The rearrangement of (chloromethyl)silanes induced by Lewis acids [143] (Eq. 36a) and the acid catalyzed rearrangement of silylcarbinols [144,145] (Eq. 36b) belong to this category.

$$Me_3SiCH_2Cl \xrightarrow{AlCl_3} Me_2\overset{Cl}{\underset{|}{Si}}-CH_2-CH_3 \qquad (36a)$$

$$Me_2\overset{Ph}{\underset{|}{Si}}-CH_2OH \xrightarrow{HF} Me_2\overset{F}{\underset{|}{Si}}-CH_2Ph \qquad (36b)$$

The stereochemical course of these reactions can be rationalized on the basis of Woodward-Hoffman rules. The reaction proceeds as a 1,2-sigmatropic suprafacial process, and since the highest occupied molecular orbital

(HOMO) is of the σ + π type, retention on the migrating center is observed (Eq. 37).

b) A Free Electron Pair in the β Position to Silicon

This category of reactions is represented by the base catalyzed rearrangement of silylcarbinols [146,147]. Reaction proceeds in a fashion that is consistent with the requirement of conservation of orbital symmetry with inversion at silicon [148,149] (Eq. 38).

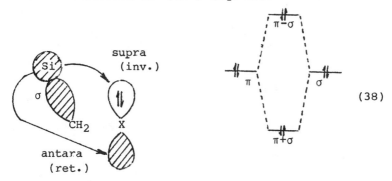

In principle, the antarafacial course of reaction with retention at silicon is also possible, but steric factors probably favor the suprafacial migration. The reaction proceeds easily since the products are stabilized by the formation of a very stable Si-O bond.

The same stereochemical course would also be followed by the radical rearrangement of α-peroxides [150] (Eq. 39).

$$R_3Si-\underset{R'}{CH}-OOH \xrightarrow{\Delta} R_3Si-O-\underset{R'}{CH}-OH \qquad (39)$$

The photochemical rearrangement of α-silylketones [151] also belongs to the same category of reactions (Eq. 40).

$$R_3SiCOR' \xrightarrow{h\nu} [R_3Si-O-\ddot{C}R'] \tag{40}$$

The fate of a carbenoid intermediate depends strongly on the basicity of the environment. In nonbasic solvents a resonance stabilization of the β carbenoid center takes place and the following reaction is observed (Eq. 41).

$$R_3SiCOR' \xrightarrow{R''OH} R_3SiOR + R'CH(OR'')_2 \tag{41}$$

On the other hand, in the presence of base the reaction proceeds as follows (Eq. 42).

$$R_3SiCOR' \xrightarrow{R''O^-} R_3Si-O-\underset{\underset{OR''}{|}}{\overset{-}{C}}-R' \xrightarrow{H^+} R_3Si-O-\underset{\underset{OR''}{|}}{CHR'} \tag{42}$$

The photochemical decomposition of bis(trialkylsilyl) ketones proceeds in a different way [152] (Eq. 43).

$$R_3SiCOSiR_3 \xrightarrow{h\nu} R_3Si-SiR_3 + CO \tag{43}$$

The photochemical fragmentation of polysilanes [153] proceeds analogously (Eq. 44).

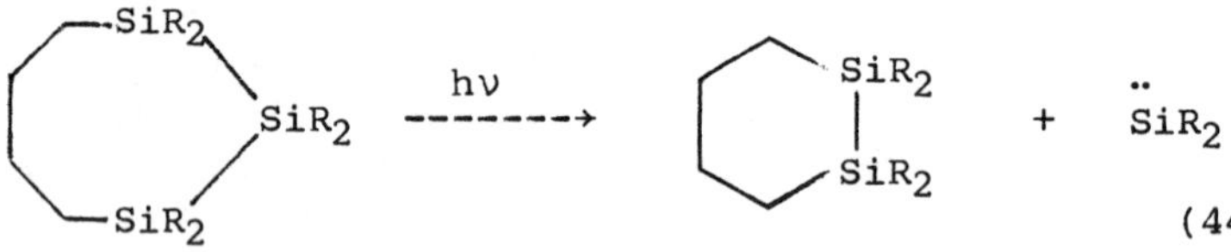

(44)

Both reactions correspond to chelatotropic reactions in the Woodward-Hoffmann classification and proceed by both linear and nonlinear mechanisms.

The thermal decarbonylation of bis(trialkylsilyl) ketones is symmetry forbidden. Allowed, however, is a thermal decarbonylation of silanecarboxylic acids [154]. The latter reaction can also be induced by the presence of bases [155].

c) A Positive Charge in the β Position to Silicon

The orbital correlation diagram demonstrates that the frontier orbital is the π + σ type (Eq. 45).

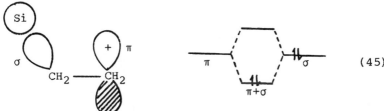

(45)

Interaction of the corresponding σ and π orbitals leads to a partial charge transfer from the Si-C bond to the positively charged β carbon, which is accompanied by a partial double bond character of the central C-C bond, and a partial positive charge appears on the silicon. This explains the well known tendency to split the Si-C bond with the elimination of ethene (β-elimination [156-158]). Similarly, β-silylketones also decompose in the presence of bases [159] (Eq. 46).

$$R_3Si-CH_2-CO(R') \xrightarrow{B^-} R_3SiB + CH_3COR' \quad (46)$$

Yet another type of interesting thermal rearrangement [160] appears in these systems (Eq. 47).

$$R_3Si-CH_2COR \xrightarrow{\Delta} R_3Si-O-C(=CH_2)-R' \quad (47)$$

This rearrangement can be characterized as a 1,3-sigmatropic reaction proceeding either antarafacially with retention at silicon or suprafacially with inversion. To distinguish between these two alternatives, it would be desirable to study this reaction in cyclic systems.

Because of partial delocalization of positive charge from carbon to silicon, it is obvious that it has no meaning to ask whether the transition state in the solvolysis of β-haloethyl or β-hydroxyethylsilanes resembles a carbonium or a silicenium ion. Rather, it would be

interesting to decide on the basis of the Traylor test
[161, 162] whether the stabilization of the β-carbonium
center proceeds vertically (hyperconjugation) or by neighboring group participation (silacyclopropenium ion). As
is indicated by the results of ab initio calculations, the
silyl group is much more effective in stabilizing a β-
cationic center than is an alkyl group [163].

The goal of the above section has been to attempt to
unify theoretical explanations of the observed reactivity
of carbon-functional silanes rather than to provide an
extensive survey of all results from the field. An
interested reader can find a much more complete survey in
the chapter by J. Pola (Chapter 2).

4.8 THE CHEMISTRY OF MULTIPLE BONDED SILICON

The tendency of silicon to avoid the formation of
multiple bonds is still one of the challenging problems
for valence theory. The first attempts to explain the
instability of silicon analogues of unsaturated compounds
were based on the Pitzer idea of repulsion of occupied
orbitals and the lower values of overlap integrals for the
elements of the third period [164,165]. On the basis of
these theoretical conclusions the existence of such compounds was assumed to be impossible [166]. The first convincing proof of the existence of derivatives of
silaethene as reactive intermediates during the pyrolysis
of silacyclobutanes was given by Gusel'nikov [167,168].
This work stimulated further effort to detect these compounds, and their formation was confirmed in numerous
reactions. In addition to intermediates with a Si=C bond,
the existence of compounds with Si=O and Si=N bonds was
also detected. Nevertheless, the most frequently studied
molecules of this kind are still the derivatives of
silaethene. Their transient existence was proved in many
reactions, e.g., in the pyrolysis of silacyclobutanes
[167,168] (Eq. 48),

$$\text{Me}_2\text{Si}\text{-cyclobutane} \xrightarrow[-C_2H_4]{>400°} [\text{Me}_2\text{Si}=\text{CH}_2] \quad (48)$$

the photolysis of silyldiazoalkanes [169] (Eq. 49),

THEORETICAL ASPECTS OF BONDING

$$Me_3Si-CH=N=N \xrightarrow{h\nu} [Me_3Si-\ddot{C}H] \dashrightarrow [Me_2Si=CHMe] \quad (49)$$

the photolysis of acylpolysilanes [170] (Eq. 50),

$$R_3Si-SiR_2-COR' \xrightarrow{h\nu} [R_3Si-SiR_2-O-\ddot{C}R'] \quad (50)$$
$$\xrightarrow{h\nu} [R_2Si=\underset{\underset{R'}{|}}{C}-OSiR_3]$$

and in the Jones synthesis [171] (Eq. 51),

$$\underset{\underset{Cl}{|}}{Me_2Si}-CH=CH_2 \xrightarrow[-78°]{tBuLi} [\underset{\underset{Cl}{|}}{Me_2Si}-\underset{\underset{Li}{|}}{CH}-CH_2tBu] \xrightarrow{-LiCl} [Me_2Si=CH-CH_2tBu]$$
$$(51)$$

Intermediate with the structure of silaethene derivatives are very reactive, and the proof of their existence is frequently based only on an analysis of the structure of the products of the reaction of a silaethene derivative with some trapping agent. As is documented by the following examples, the derivatives of silaethene act as electrophilic reagents [172] (Eq. 52).

$$[Me_2Si=CH_2] \begin{array}{l} \xrightarrow{CH_3CN} Me_3Si-CH_2CN \\ \xrightarrow{PhOH} Me_3Si-CH_2OPh \\ \xrightarrow{PhNH_2} Me_3Si-CH_2NHPh \end{array} \quad (52)$$

The fact that silaethene and triphenylphosphinmethylid react similarly with ketones tends to confirm the dipolar character of the Si=C bond (Eq. 53).

$$[Me_2Si=CH_2] \xrightarrow{Ph_2CO} Ph_2C=CH_2 + [Me_2Si=O] \quad (53)$$

This reaction is also interesting because its product is a reactive intermediate with an Si=O bond [172]. Such compounds also appear in the pyrolysis of cyclocarbosiloxanes and cyclosiloxanes [168] (Eq. 54).

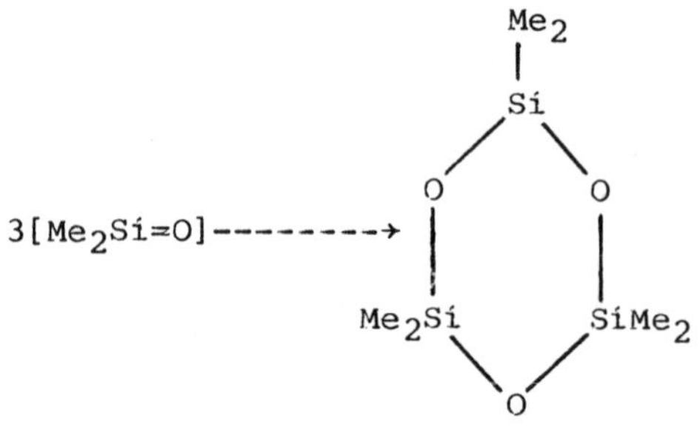

$$3[Me_2Si=O] \dashrightarrow \underset{[D_3]}{\begin{array}{c} Me_2 \\ | \\ Si \\ O \diagup \diagdown O \\ | | \\ Me_2Si SiMe_2 \\ \diagdown O \diagup \end{array}} \qquad (54a)$$

$$D_n \xrightarrow{\Delta} D_{n-1} + [Me_2Si=O] \qquad (54b)$$

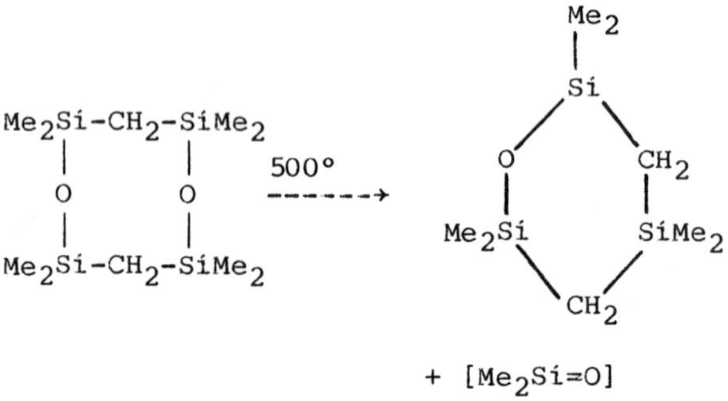

$$\begin{array}{c} Me_2Si-CH_2-SiMe_2 \\ | | \\ O O \\ | | \\ Me_2Si-CH_2-SiMe_2 \end{array} \xrightarrow{500°} \begin{array}{c} Me_2 \\ | \\ Si \\ O \diagup \diagdown CH_2 \\ | | \\ Me_2Si SiMe_2 \\ \diagdown CH_2 \diagup \end{array} \qquad (54c)$$

$$+ [Me_2Si=O]$$

Intermediates with Si=N bonds were detected in the photolysis of silylazides [173] (Eq. 55).

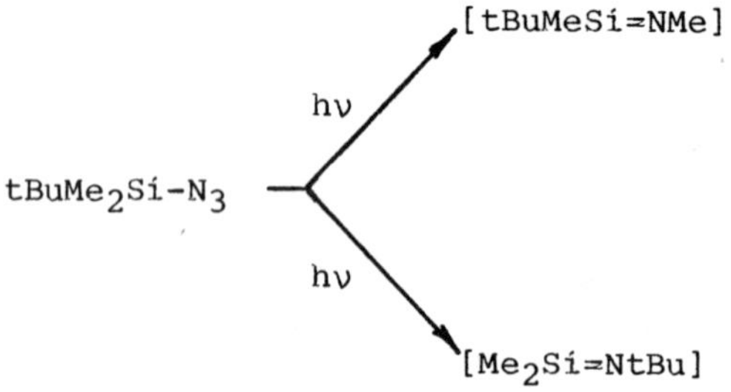

$$tBuMe_2Si-N_3 \begin{array}{c} \nearrow^{h\nu} [tBuMeSi=NMe] \\ \\ \searrow_{h\nu} [Me_2Si=NtBu] \end{array} \qquad (55)$$

The low stability and the high reactivity of all of these intermediates have stimulated the effort of theoretical chemists to take an increasing interest in elucidating the structures of these intermediates. In this respect the most frequently studied molecule is silaethene, for which calculations were performed on both semiempirical [174,175] and ab initio levels [176-181]. One of the principal tasks of such calculations consists of finding the most stable isomer on the potential energy hypersurface of the $SiCH_4$ species. Gordon [182] and Schäfer [183,184] report on the basis of ab initio calculations that the most stable isomer is the singlet state of methylsilylene (CH_3-Si-H). More recent calculations by Köhler and Lischka [185] and by Schäfer [186] demonstrate, however, that after including the correlation energy corrections, both molecules become essentially degenerate.

Another problem represents the characterization of the structure of silaethene. Here it is important to decide whether the ground state is represented by the closed shell singlet, as described by a zwitter-ionic structure, or by an open shell biradical triplet (Eq. 56).

$$\overset{+}{SiH_2}----\overset{-}{CH_2} <----> SiH_2{=}CH_2 <----> \overset{\cdot}{SiH_2}----\overset{\cdot}{CH_2} \quad (56)$$

Despite the fact that Strausz and coworkers reported in one of the first ab initio studies of this molecule that the ground state is the pyramidal open shell triplet [179], the majority of authors now agree that the silaethene structure is better described by the zwitter-ionic structure [175,178,181]. According to recent SCF studies, the triplet state lies roughly 50-60 kJ/mol above the ground state singlet. After the inclusion of CI this difference increases to approximately 140 kJ/mol [178,187].

The ground state equilibrium geometry of silaethene is now thought to be planar, with the Si-C bond length varying with the quality of the basis set or quantum chemical treatment (SCF + eventual inclusion of correlation energy) from 0.164 nm [179] to 0.172 nm [178]. All of these values, however, are much lower than the experimental result (0.183 nm) obtained recently by neutron diffraction studies of 1,1-dimethylsilaethene stabilized in an argon matrix [188]. This discrepancy between theory and experiment is probably too large to be attributed to the change of substituents on silicon. Moreover, in a recent ab initio study (4-31 G basis set) Morokuma determined the equilibrium geometry of 1,1-dimethylsilaethene and obtained a Si=C bond length of 0.169 nm [180].

It is difficult to find the true cause of such discrepancies. In addition to possible objections against the sufficiency of the SCF approximation, another important factor in correlating theoretical and experimental data on these highly unstable derivatives is the reliability of the experimental data. The theoretical interpretation of the IR spectra of silaethene may serve as an example to illustrate the uncertain reliability of experimental characteristics. Schlegel and Wolfe [177] calculated in ab initio studies that the vibration frequency ν_{SiC} is 1490 cm^{-1}, which they compared with the experimental value of ν_{SiC} = 1407 cm^{-1} reported by Barton [189] for 1,1-dimethylsilaethene stabilized in an argon matrix at 77 K. This experimental value was questioned, however, by Maltsev and coworkers [190] who report the reactivity of 1,1-dimethylsilaethene at 77 K to be too high for stabilization in an argon matrix. Their corrected value of ν_{SiC} = 1003 cm^{-1} was determined at 10 K. Similarly, Chapman and coworkers [169] report that for the trapping of 1,1,2-trimethylsilaethene it is necessary to work at 8 K. Also, for the detection of PE spectra of the parent silaethene, recently described by Bock [191], it was necessary to lower the temperature below 10 K. On the other hand, the frequency of ν_{SiC} determined by Maltsev and also recently by Meyer [192] is in very nice agreement with the theoretical value of ν_{SiC} = 1000 cm^{-1} calculated by Morokuma [180].

Morokuma also studied the relative stability of all three isomeric forms of 1,1-dimethylsilaethene (Eq. 57).

$$\mathrm{Me\ddot{S}i\text{-}CH_2Me} \xleftarrow{\ \ \ \ \ } Me_2Si\text{=}CH_2 \xleftarrow{\ \ \ \ \ } Me_2SiH\text{-}\ddot{C}H \qquad (57)$$

In contrast to the parent silaethene, where the most stable isomer corresponds to singlet methylsilylene [182-184], the energetically most stable form of the C_3SiH_8 isomers represents 1,1-dimethylsilaethene. This result demonstrates that proper substitution may stabilize compounds with Si=C bonds. Recently, stable derivatives of silaethene were prepared by the photolysis of properly substituted acylpolysilanes [193]. Especially effective in stabilizing the Si=C bond are bulky substituents such as an adamantyl group (Eq. 58).

$(Me_3Si)_3Si-CO-adamantyl \xrightarrow{h\nu}$ (58)

$$Me_3Si\diagdown\hspace{-0.3em}\underset{Me_3Si\diagup}{Si=C}\hspace{-0.3em}\diagup^{OSiMe_3}_{adamantyl}$$

The calculated rotational barrier for cis-trans isomerization of the Si=C bond amounts to 200 kJ/mol in 1,1-dimethylsilaethene [180], whereas in ethene it is approximately 275 kJ/mol. This demonstrates that the π bond in silaethene is much weaker than in ethene. Similar conclusions also follow from semiempirical calculations by Dewar [175].

Among the other multiple bonded silicon compounds, increasing attention is being devoted to derivatives of disilene. The formation of these relatively stable compounds was first detected by West [194] and also recently by a group of Japanese workers [195]. The structure of the parent disilene $H_2Si=SiH_2$ was investigated theoretically by several groups [176,196-199]. According to recent calculations, ground state disilene is planar with an Si=Si bond length of 0.212 nm (at the SCF level) [198]. Inclusion of correlation energy leads to a slight elongation of this bond to 0.217 nm [185]. Also reported is a comparison of the relative stability of Si_2H_4 isomers. At the SCF level of calculations the most stable isomer is predicted to be singlet silylsilylene, but the CEPA correlation correction makes both disilene and silylsilylene energetically nearly equivalent [199].

Although compounds containing Si=C and Si=Si bond have been studied in increasing numbers, much less attention has been devoted to other multiple bonded silicon intermediates. Thus, e.g., Kutzelnigg [200] described the equilibrium structure and force constants of the hypothetical silanone $H_2Si=O$. Its ground state corresponds to a highly polar zwitter-ionic structure. The isomeric hydroxysilylene was predicted to be more stable by about 8 kJ/mol [201].

The structure of intermediates containing a silicon-carbon triple bond was recently investigated by Pople and Gordon [202].

Among other compounds containing multiple bonded silicon, attention has been devoted to silabenzene and its derivatives. Their existence as intermediates was recently proved in the course of several reactions [203-205], and their structure was characterized using UV and IR spectroscopy in an argon matrix at 10 K [206]. Bock and coworkers also described the PE spectrum of silabenzene generated by pyrolysis of 1-allyl-1-sila-2,4-cyclohexadiene [207].

In addition to these experimental reports, silabenzene has also been studied theoretically by semiempirical [175] and ab initio methods [208,209]. The results of these calculations suggest that silabenzene should have some aromatic character. Dewar reports on the basis of MINDO/3 calculations that the resonance energy of silabenzene is approximately 30 kJ/mol [175]. From the comparison of the relative stability of C_5H_6M isomers, it follows (Eq. 59) that the reaction of silacarbene (G) to silabenzene (H) in the singlet state is approximately 120 KJ/mol less exothermic than for its carbon analogue. On the other hand, the reverse reaction is thermodynamically more favorable in the triplet state by about 300 kJ/mol.

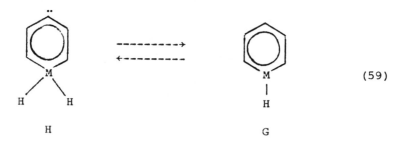

(59)

2.9 CONCLUSIONS

In the above chapter we have attempted to elucidate the physical meaning of such concepts as bonding utilization of d orbitals, hyperconjugation, hybridization, etc., in terms of valence theory. This analysis, supported by the results of numerous theoretical studies, suggests that the original ideas about the necessity of bonding utilization of d orbitals must be modified, since with the aid of ab initio quantum chemical methods it is possible to reproduce not only qualitatively but also quantitatively the numerous structural characteristics of organosilicon compounds even without the inclusion of d orbitals into the basis set. Thus, it becomes evident that the concept of $(p-d)_\pi$ interaction can be considered only as a certain simple mnemotechnic aid in explaining the electron accepting behavior of silicon.

Similar conclusions also hold for hyperconjugation. The phenomenon of hyperconjugation does not represent a new specific mechanism of intramolecular interaction; instead, it appears only as a consequence of the effort to preserve the classical picture of localized chemical bonds even where it is, strictly speaking, impossible. The numerous conclusions about the increased role of hyperconjugation in organosilicon chemistry imply therefore only that the classical picture of chemical bonds holds to a smaller extent in organosilicon chemistry than in classical organic chemistry. The deviations are frequently so remarkable that in e.g., the series of alicyclic polysilanes, one speaks about aromatic behavior.

Thus, neither the concept of $(p-d)_\pi$ interaction nor hyperconjugation can be given physical meaning in the sense that they represent existing mechanisms of intramolecular interaction. The question therefore arises of what is the real cause of the observed differences between organic and organosilicon chemistry. From the accumulated theoretical results it becomes evident that the decisive factor that differentiates the two elements is the greater polarizability of silicon, which manifests itself in the greater ability of electron distribution to accomodate the changes induced by external physical or chemical perturbations. The increased polarizability of silicon thus appears, from the theoretical point of view, to be the best means of characterizing the differences between organic and organosilicon chemistry.

ACKNOWLEDGMENT

The author wishes to express his gratitude to Dr. J. Pecka (Department of Organic Chemistry, Charles University, Prague) for critically reading the manuscript and for helpful suggestions.

REFERENCES

1. Eaborn C.: Organosilicon compounds. Butterworths Ltd., London, 1960.
2. Pauling L.: Nature of the Chemical Bond. Cornell Univ. Press, New York, 1939.
3. Craig D.P., Maccoll A., Nyholm R., Orgel L.E., Sutton L.E.: J. Chem. Soc. 332 (1954).
4. Craig D.P., Maccoll A., Nyholm R.S., Sutton L.E.: J. Chem. Soc. 354 (1954).
5. Jaffé H.H.: J. Phys. Chem. 58, 185 (1954).
6. Mitchell K.A.R.: Chem. Rev. 69, 157 (1969).
7. Ebsworth E.A.W.: Organometallic Compounds of the Group IVb Elements, (MacDiarmid A.G., ed.), Marcel Dekker Inc., New York, 1968.
8. Brill T.B.: J. Chem. Educ. 50, 392 (1973).
9. Lucken E.A.: Structure and Bonding, Vol. 6, pp. 1-29, Springer, Berlin, Heidelberg, 1969.
10. Jørgensen C.K. Structure and Bonding, Vol. 6, pp. 94-115, Springer, Berlin, Heidelberg, 1969.
11. Pitt C.G.: J. Organometal. Chem. 61, 49 (1973).
12. Born M., Oppenheimer R.J.: Ann. Phys. 84, 457 (1927).
13. Seyferth D., Pudvin J.J.: Chem. Tech. 231, April 1981.
14. Pauling L.: J. Amer. Chem. Soc. 53, 1367 (1931).
15. Kimball J.R.: J. Chem. Phys. 8, 188 (1940).
16. Maccoll A.: Trans. Faraday Soc. 46, 369 (1950).
17. Coulson C.A., Moffitt W.E.: Phil. Mag. 40, 1 (1949).
18. Ponec R.: Unpublished results.
19. Slater J.: Phys. Rev. 36, 57 (1930).
20. Boys S.F.: Proc. Roy. Soc. A200, 542 (1950).
21. Craig D.P., Magnusson E.A.: J. Chem. Soc. 4895 (1956).
22. Boer F.P., Lipscomb W.: J. Chem Phys. 50, 989 (1969).
23. Roos B., Siegbahn P.: Theor. Chim. Acta 17, 199 (1970).
24. Collins J.B., von Schleyer P.R., Binkley J.S., Pople J.A.: J. Chem. Phys. 64, 5142 (1976).
25. Webster B.C.: J. Chem. Soc. A, 2909 (1968).
26. Gillespie J.R.: J. Chem. Soc. 1002 (1952).
27. Moccia R.: J. Chem. Phys. 40, 2164 (1962).
28. Hatano J.: Chem. Phys. Lett. 56, 314 (1978).
29. Roothaan J.C.C.: Rev. Mod. Phys. 23, 69 (1951).

30. Daudel R., Poirier R.A., Goddard J.D., Csizmadia I.G.: Int. J. Quant. Chem. 15, 261 (1979).
31. Mezey P.G., Yates K., Theodorakopoulos T.G., Csizmadia I.G.: Int. J. Quant. Chem. 12, 247 (1977).
32. Huzinaga S., Arnau C.: J. Chem. Phys. 53, 348 (1970).
33. Veillard A.: Theor. Chim. Acta 12, 405 (1968).
34. Oberhammer H., Boggs J.E.: J. Amer. Chem. Soc. 102, 7241 (1980).
35. Ratner M.A., Sabin J.R.: J. Amer. Chem. Soc. 93, 3542 (1971).
36. Ratner M.A., Sabin J.R.: J. Amer. Chem. Soc. 99, 3954 (1977).
37. Rodwell R.W.: J. Amer. Chem. Soc. 100, 7209 (1978).
38. Chvalovský V.: Main Lecture, 1st International Symp. on Organosilicon Chem., Prague, September 6-9, 1965, p. 231.
39. Coulson C.A.: Nature 221, 1107 (1969).
40. Mulliken R.S.: J. Chem. Phys. 7, 339 (1939).
41. Mulliken R.S., Rieke C.A., Brown W.C.: J. Amer. Chem. Soc. 63, 41 (1941).
42. Dewar M.J.S.: Hyperconjugation, The Ronald Press Comp., New York 1962.
43. Nesmeyanov N.A., Lutchenko F.: Dokl. Akad. Nauk 59, 707 (1948).
44. Petrov A.D., Egorov Yu.P., Mironov V.F., Nikishkin G.I., Bulgakova A.A.: Izv. Akad. Nauk. USSR, Ser. Khim. 50(1956).
45. Leites L.A.: Izv. Akad. Nauk. USSR, Ser. Khim .1525 (1963).
46. Egorov Yu.P., Leites L.A., Tolstikova N.G., Tschernyschev E.A.: Izv. Akad. Nauk USSR, Ser. Khim. 445 (1961).
47. Pitt C.G.: Chem. Commun. 816 (1971).
48. Heilbronner E., Hornung W., Bock H., Alt H.: Angew. Chem. 81, 537 (1969).
49. Bock H., Ensslin W.: Angew. Chem. 83, 435 (1971).
50. Mollére P., Bock H., Becker G., Fritz G.: J. Organometal. Chem. 46, 89 (1972).
51. Bock H., Mollére P., Becker G., Fritz G.: J. Organometal. Chem. 61, 113 (1973).
52. Bock H., Pitt C.G.: Chem. Commun. 5, 28 (1972).
53. Mollére P., Bock H., Becker G., Fritz G.: J. Organometal. Chem. 61, 127 (1973).
54. Bock H., Ensslin W., Fehér F., Freund R.: J. Amer. Chem. Soc. 98, 668 (1976).
55. Bock H., Wittel K., Veith M., Wiberg N.: J. Amer. Chem. Soc. 98, 109 (1976).

56. Weidner U., Schweig A.: J. Organometal. Chem. 39, 261 (1972).
57. Schweig A., Weidner U., Manuel G.: Angew. Chem. 84, 899 (1972).
58. Schmidt H., Schweig A.: Tetrahedron Lett. 1973, 981.
59. Schmidt H., Schweig A.: Angew. Chem. 85, 299 (1973).
60. Ponec R., Chvalovský V.: Coll. Czech. Chem. Commun. 38, 3845 (1973).
61. Ponec R., Tschernyschev E.A., Tolstikova E.A., Chvalovský V.: Coll. Czech. Chem. Commun. 41, 2714 (1976).
62. Ponec R., Chvalovský V., Tschernyschev E.A., Komarenkova N.G., Bashkirova S.A.: Coll. Czech. Chem. Commun. 39, 1177 (1974).
63. Ponec R., Chvalovský V.: Coll. Czech. Chem. Commun. 39, 1185 (1974).
64. Ponec R., Chvalovský V., Tschernyschev E.A., Shchepinov, S.A., Krasnova T.L., Coll. Czech. Chem. Commun. 39, 1313 (1974).
65. Hoffmann R., Radom L., Pople J.A., von Schleyer P.R., Hehre W.J., Salem L.: J. Amer. Chem. Soc. 94, 6221 (1972).
66. Leites L.A., Finkelstein E.S., Vdovin V.M., Nametkin N.S.: Izv. Akad. Nauk USSR, Ser. Khim. 1305 (1965).
67. Leites L.A., Gar T.K., Mironov V.F.: Dokl. Akad. Nauk USSR 158, 400 (1964).
68. Pitt C.G.: J. Organometal. Chem. 23, C35 (1970).
69. Eaborn C., Jackson P.M., Taylor R.: J. Chem. Soc. B 613 (1966).
70. Eaborn C., Pande K.C.: J. Chem. Soc. 1566 (1960).
71. Berwin H.J.: Chem. Commun. 237 (1972).
72. Goodman L., Konstam A.H., Sommer L.H.: J. Amer. Chem. Soc. 87, 1012 (1965).
73. West R.: J. Organometal. Chem. 3, 315 (1965).
74. Harnisch F.D., West R.: Inorg. Chem. 2, 1082 (1963).
75. Agolini F., Klemenko S., Csizmadia I.G., Yates K.: Spectrochim. Acta 424, 169 (1968).
76. Ramsay B.G., Brook A.G., Bassindale R.A., Bock H.: J. Organometal. Chem. 74, C41 (1974).
77. Curran C., Witucki R., McCusker P.: J. Amer. Chem. Soc. 72, 4471 (1950).
78. Dejmek L., Ponec R., Chvalovský V.: Coll. Czech. Chem. Commun. 45, 3518 (1980).
79. Pola J., Chvalovský V.: Coll. Czech. Chem. Commun. 41, 581 (1976).
80. Pola J., Chvalovský V.: Coll. Czech. Chem. Commun. 43, 746 (1978).

81. Coulson C.A.: Trans. Faraday Soc. 38, 433 (1942).
82. Lenard-Jones J.: Proc. Roy. Soc. A198, 14 (1949).
83. Millié P., Lévy B., Berthier G.: in Localisation and Delocalisation in Quantum Chemistry, Vol. 1 (Chalvet O., Diner S., Daudel R., Malrieu J.P., eds.) D. Reidel Publ. Comp., Dordrecht, 1975.
84. Dewar M.J.S.: MO Theory of Organic Chemistry, McGraw-Hill, New York, 1969.
85. Ponec R.: J. Mol. Struct. (THEOCHEM) 86, 285 (1982).
86. Diner S., Malrieu J.P., Claverie P.: Theor. Chim. Acta 13, 1 (1969).
87. Aslangul C., Kottis P., Constanciel R., Daudel R.: Adv. Quant. Chem. 6, 93 (1972).
88. Polák R.: Int. J. Quant. Chem.: 6, 1077 (1972).
89. Polák R.: Theor. Chim. Acta 14, 163 (1969).
89a. West R.: Pure Appl. Chem. 54, 1041 (1982).
90. Sommer L.H., Gold J.R., Goldberg G.M., Marans N.S.: J. Amer. Chem. Soc. 71, 1509 (1949).
91. Voronkov M.G., Feshin V.P., Romanenko L.S., Pola J., Chvalovský V.: Coll. Czech. Chem. Commun. 41, 2718 (1976).
92. Whitmore F., Sommer L.H.: J. Amer. Chem. Soc. 68, 481 (1946).
93. Ponec R.: Theor. Chim. Acta: 59, 629 (1981).
94. Hinze J., Jaffé H.H.: J. Amer. Chem. Soc. 84, 540 (1962).
95. Pople J.A., Santry D.P., Segal G.: J. Chem. Phys. 43, 5129 (1965).
96. Gilman H., Dunn G.E.: J. Amer. Chem. Soc. 73, 3404 (1951).
97. Schott G., Gutschick G.: Z. anorg. allg. Chem. 325, 175 (1963).
98. Schott G., Harzdorf G.: Z. anorg. allg. Chem. 306, 180 (1960).
99. Seyferth D., Damrauer R., Pui Mui Yick J., Jula T.F.: J. Amer. Chem. Soc. 90, 2944 (1968).
100. Bott R.W., Eaborn C., Pande K.G., Swadle T.W.: J. Chem. Soc. 1217 (1962).
101. Pola J., Bažant V., Chvalovský V.: Coll. Czech. Chem. Commun. 37, 3885 (1972).
102. Tschernyschev E.A., Tolstikova N.G.: Izv. Akad. Nauk USSR, Ser. Khim. 455 (1961).
103. Roberts J.D., McElhill E.A., Armstrong R.: J. Amer. Chem. Soc. 71, 2923 (1949).
104. Benkeser R.A., Krysiak H.R.: J. Amer. Chem. Soc. 75, 2421 (1953).
105. Benkeser R.A., De Boer H.R., Robinson R.E., Sauve D.M.: J. Amer. Chem Soc. 78, 682 (1956).

106. Roberts J.D., Regan C.: J. Amer. Chem. Soc. 75, 4102 (1953).
107. Mareš F., Plzák Z., Hetflejš J., Chvalovský V.: Coll. Czech. Chem. Commun. 36, 2957 (1971).
108. Vo-Kim-Yen, Papoušková Z., Schraml J., Chvalovský V.: Coll. Czech. Chem. Commun. 38, 3167 (1973).
109. Fialová M., Bažant V., Chvalovský V.: Coll. Czech. Commun. 38, 3837 (1973).
110. Wang T.J., Van Dyke C.H.: Inorg. Chem. 6, 1741 (1967).
111. Gibbon G.A., Wang T.J., Van Dyke C.H.: Inorg. Chem. 6, 1989 (1967).
112. Voronkov M.G., Feshin V.P., Mironov V.F., Mikhailants S.S., Gar T.K.: Zh. Obshch. Khim. 41, 2211 (1971).
113. Pola J., Schraml J., Chvalovský V.: Coll. Czech. Chem. Commun. 38, 3158 (1973).
114. Ponec R., Chvalovský V.: Coll. Czech. Chem. Commun. 40, 2309 (1975).
115. Ponec R., Chvalovský V.: Coll. Czech. Chem. Commun. 40, 2480 (1975).
116. Eaborn C., Parker H.S.: J. Chem. Soc. 939 (1954).
117. Cook M.A., Eaborn C., Walton D.R.M.: J. Organometal. Chem. 24, 293 (1970).
118. Jarvie A.W.P.: Organometal. Chem. Rev. A6, 153 (1970).
119. Ponec R., Dejmek L., Chvalovský V.: J. Organometal. Chem. 197, 31 (1980).
120. Libitt L., Hoffmann R.: J. Amer. Chem. Soc. 96, 1370 (1974).
121. Hopkinson A., Lien M.H.: J. Org. Chem. 46, 998 (1981).
122. Radom L., Hehre W., Pople J.A.: J. Amer. Chem Soc. 94, 2371 (1972).
123. Pople J.A., in: The Word of Quantum Chemistry. (Daudel R., Pullmann B., eds.), D. Reidel, Dordrecht 1974.
124. Ponec R., Dejmek L., Chvalovský V.: Coll. Czech. Chem. Commun. 45, 2895 (1980).
125. Dejmek L., Ponec R., Chvalovský V.: Coll. Czech. Chem. Commun. 45, 3510 (1980).
126. Field H.F.: Accounts. Chem. Res. 1, 42 (1968).
127. Beauchamp J.L.: Ann. Rep. Phys. Chem. 22, 527 (1971).
128. Ponec R., Kučera J., Chvalovský V.: Coll. Czech. Chem. Commun. 48, 1602 (1983).
129. Ponec R., Kučera J.: Chem. Scripta, 22, 152 (1983).
130. Boys F., Bernardi F.: Mol. Phys. 19, 553 (1970).
131. Pola J., Voronkov M.G., Ponec R., Chvalovský V., Kishko Yu.: Zh. Obshch. Khim., in press.
132. Corriu R.P.J., Henner M.: J. Organometal. Chem. 74, 1 (1974).

133. Corey J.Y., Gust D., Mislow K.: J. Organometal. Chem. 101, C7 (1975).
134. Corey J.Y.: J. Amer. Chem. Soc 97, 3237 (1975).
135. Lambert J.B., Hsiang-Ning-Sun: J. Amer. Chem. Soc. 98, 5611 (1976).
136. Barton T.J., Howland A.K., Tully C.R.: J. Amer. Chem. Soc. 97, 5695 (1976).
137. Litzow M.R., Spanding L.T., in: Mass Spectrometry of Inorganic and Organometallic Compounds, Phys. Inorg. Chem. Monograph 2, (Lappert M.F., ed.), Elsevier, New York 1973.
138. Apeloig Y., von Schleyer P.R.: Tetrahedron Lett., 1977, 4647.
139. Murphy M.K., Beauchamp L.J.: J. Amer. Chem. Soc. 99, 2085 (1977).
140. von Uhlig E., Hipler B., Muller P.Z.: Z. anorg. allg. Chem. 442, 11 (1978).
141. Evans A.G., Hamid M.A., Rees H.N.: J. Chem. Soc. 1110 (1971).
142. Olah G., Hunadi R.J.: J. Amer. Chem. Soc. 102, 6989 (1980).
143. Whitmore F.C., Sommer L.H., Gould J.R.: J. Amer. Chem. Soc. 69, 1976 (1947).
144. Brook A.G., Pannell K.H., Le Grow G.E., Sheets J.J.: J. Organometal. Chem. 2, 491 (1964).
145. Pannell K.H.: Diss. Abstr. 4505 B (1968).
146. Brook A.G.: Pure Appl. Chem. 13, 215 (1966).
147. Brook A.G., Le Grow L.E., Mac Ree D.N.: Can. J. Chem. 45, 239 (1967).
148. Brook A.G., Warner C.M., Limburg W.W.: Can. J. Chem. 45, 1231 (1967).
149. Biernbaum M.S., Mosher H.S.: J. Amer. Chem. Soc. 93, 6221 (1971).
150. Eisch J.J., Husk G.R.: J. Org. Chem. 29, 254 (1964).
151. Brook A.G., Duff M.J.: J. Amer. Chem. Soc. 89, 454 (1967).
152. Brook A.G., Peddle G.J.D.: J. Organometal. Chem. 5, 106 (1966).
153. Ramsay B.G.: J. Organometal. Chem. 67, C67 (1974).
154. Brook A.G.; J. Amer. Chem. Soc. 77, 4827 (1955).
155. Brook A.G., Gilman H.: J. Amer. Chem. Soc. 77, 2322 (1955).
156. Sommer L.H., Goldberg G.M., Dorfmann E., Whitmore F.C.: J. Amer. Chem. Soc. 68, 1083 (1946).
157. Sommer L.H., Bailey D.L., Whitmore F.C.: J. Amer. Chem. Soc. 70, 2869 (1948).

158. Sommer L.H., Bailey D.L., Goldberg G.M., Buch C.E., Bye T.S., Evans F.J., Whitmore F.C.: J. Amer. Chem. Soc. 76, 1613 (1954).
159. Gold R.J., Sommer L.H., Whitmore F.C.: J. Amer. Chem. Soc. 70, 2874 (1948).
160. Brook A.G., Mac Rae D.M., Limburg W.W.: J. Amer. Chem. Soc. 89, 5493 (1967).
161. Hanstein W., Berwin H.J., Traylor T.G.: J. Amer. Chem. Soc. 92, 7476 (1970).
162. Traylor T.G., Hanstein W., Berwin H.J., Clinton N.A., Brown R.S.: J. Amer. Chem. Soc. 93, 5715 (1971).
163. Eaborn C., Fleichtmar F., Horn M., Murrel J.N.: J. Organometal. Chem. 77, 39 (1974).
164. Ballard R.E.: Organometallic Chem. Rev. 2. Elsevier, Amsterdam 1976.
165. Pitzer K.S.: J. Amer. Chem. Soc. 70, 2140 (1948).
166. Dasent W.E.: Nonexistent Compounds, Marcel Dekker, New York 1965.
167. Nametkin N.S., Vdovin V.M., Guselnikov L.E., Zavyalov V.I.: Izv. Akad. Nauk USSR, Ser. Khim. 584 (1966).
168. Gusel'nikov L.E., Nametkin N.S., Vdovin V.M.: Accounts Chem. Res. 8, 18 (1975).
169. Chapman O.L., Chang C.C., Kolc J., Jung M.E., Lowe J.A., Barton T.J., Tumey M.L.: J. Amer. Chem. Soc. 98, 7845 (1976).
170. Brook A.G., Harris J.W., Lennon J., El Sheikh M.: J. Amer. Chem. Soc. 101, 83 (1979).
171. Jones R.P., Lim T.O.: J. Amer. Chem. Soc. 99, 8447 (1977).
172. Busch R.D., Golino C.M., Horner G.P., Sommer L.H.: J. Organometal. Chem. 80, 37 (1974).
173. Parker R.D., Sommer L.H.: J. Amer. Chem. Soc. 98, 618 (1976).
174. Damrauer R., Williams R.D.: J. Organometal. Chem. 66, 241 (1974).
175. Dewar M.J.S., Lo D.H., Ramsden C.A.: J. Amer. Chem. Soc. 97, 1311 (1975).
176. Blustin P.H.: J. Organometal. Chem. 105, 161 (1976).
177. Schlegel B.H., Wolfe S., Mislow K.: J. Chem. Soc. Chem. Commun. 246 (1975).
178. Hood D.M., Schäfer H.F.: J. Chem. Phys. 68, 2985 (1978).
179. Strausz P.P., Gammie L., Theodorakopoulos G., Mezey P.G., Csizmadia I.G.: J. Amer. Chem. Soc. 98, 16 (1976).

180. Hanamura M., Nagase S., Morokuma K.: Tetrahedron Lett. 1981, 1813.
181. Ahlrichs R., Heinzmann R.: J. Amer. Chem. Soc. 99, 7452 (1977).
182. Gordon M.S.: Chem. Phys. Letters 54, 9 (1978).
183. Schäfer H.F.: Accounts Chem. Res. 12, 288 (1979).
184. Goddard D.J., Yoshioka Y., Schäfer H.F.: J. Amer. Chem. Soc. 102, 7644 (1980).
185. Köhler H.J., Lischka H.: J. Amer. Chem. Soc. 104, 5884 (1982).
186. Schäfer H.F.: Accounts Chem. Res. 15, 283 (1982).
187. Vasudevan K., Grein F.: Chem. Phys. Letters 75, 75 (1980).
188. Mahaffy P.G., Gutowski R., Montgomery L.K.: J. Amer. Chem. Soc. 102, 2854 (1980).
189. Barton T.J., McIntosh L.C.: J. Chem. Soc. Chem. Commun. 861 (1972).
190. Maltsev A.K., Chabatsescu V.N., Nefedov O.M.: Izv. Akad. Nauk USSR, Ser. Khim. 1193 (1976).
191. Rosmus P., Bock H., Solouki B., Maier G., Mihm G.: Angew. Chem. 93, 616 (1981).
192. Maier G., Mihm G., Heisenauer H.P.: Angew. Chem. 93, 615 (1981).
193. Brook A.G., Nyburg G., Reynolds W.F., Poon Y.C., Yan-Min-Chang, Lee J.S., Picard J.P.: J. Amer. Chem. Soc. 103, 6750 (1981).
194. West R., Fink J.M., Michl J.: Science 214, 1343 (1981).
195. Masamune S., Hanzawa Y., Murakami S., Ballty T., Blount J.F.: J. Amer. Chem. Soc. 104, 1150 (1982).
196. Poirier R.A., Goddard J.D.: Chem. Phys. Lett. 80, 37 (1981).
197. Snyder L., Wassermann Z.R.: J. Amer. Chem. Soc. 101, 5222 (1979).
198. Lischka H., Köhler H.J.: Chem. Phys. Lett. 85, 467 (1982).
199. Daudel R., Karl R.E., Poirier R.A., Goddard J.D., Csizmadia I.G.: J. Mol. Struct. 50, 115 (1978).
200. Jaquet R., Kutzelnigg W., Staemler W.: Theor. Chim. Acta 54, 205 (1980).
201. Ponec R.: Z. Phys. Chemie, in press.
202. Gordon M.S., Pople J.A.: J. Amer. Chem. Soc. 103, 2945 (1981).
203. Barton T.J., Banasiak D.S.: J. Amer. Chem. Soc. 99, 5199 (1977).
204. Maier G., Mikun G., Reisenauer H.P.: Chem. Ber. 115, 801 (1982).

205. Barton T.J., Vuper M.: J. Amer. Chem. Soc. 103, 6789 (1981).
206. Maier G., Mikun G., Reisenauer H.P.: Angew. Chem. 92, 58 (1980).
207. Solouki B., Rosmus P., Bock H., Maier G.: Angew. Chem. 92, 56 (1980).
208. Schlegel H.B., Coleman B., Jones M. Jr.: J. Amer. Chem. Soc. 100, 6499 (1978).
209. Blustin P.H.: J. Organometal. Chem. 166, 21 (1979).

INDEX

Acidity-basicity, 70
  of organosilicon amines, 55
  of organosilicon ethers, 70, 81
  quantum chemical calculations, 271
Affinity chromatography, 23
Agents
  coupling, 10
  derivatizing, 9
  silylating, 9
AJCP method, 125
Alpha effect, 261, 265, 271
Anchimeric assistance
  heterolysis of Si-C bond, 63, 89
  ionization of $C-SO_2Cl$ bond, 60
  ionization of C-X bond, 75
  trans elimination, 76
Approximation, Born-Oppenheimer, 234
Atomic orbitals, classification of, 235

Back-bonding, 141, 213, 265
Back-donation, 141, 269
Basis set
  extent of, 248
  flexibility of, 248
Beta effect, 265
Biological activity, 1, 2, 9, 18
Bond angles
  SiOSi, 248
Bond, Si-C, 4
  cleavage of, 4, 264
Bridging interaction, 77, 78

Carbon-functional compounds
  aliphatic, 122, 142
  aromatic, 122, 196
Charge-transfer spectra, 251
Chemical reactivity, classification of, 277
Chemical shift
  ^{29}Si, general trends, 129
  ^{29}Si, theory, 132
Chromatographic purification
  of antithrombin, 24
  of carboxypeptidase, 24
  of heparin, 24
Composite material, 11, 14
Conformation, 144, 196
  effect on basicity, 274
  effect on reactivity, 53, 63
Coupling constants, 125, 142, 144, 197
CP method, 125

DEPT method, 125
Diamagnetic shielding, 133
Dihedral angle, 145
Dipole moments, 253
Direct synthesis, 2
Disilene, quantum chemical calculations of, 287
Dissociation of organosilicon
  benzoic acids, 50, 67
  carboxylic acids, 54
  cyanohydrins, 53
d orbitals
  contraction of, 246
  diffuseness of, 245
  energetic aspects of participation of, 246

d orbitals (continued)
  hybridization of, 241
  neglect of, 233, 265
  participation of, 244, 245
  at silicon, 248
  transformation properties of, 237
Double bond polarity, 195
Dual polar effects, 59
DSPI method, 125

Edman degradation, 25
Effect of substituent, 153, 264
Elasticity modulus, 18
Electron capture detection, 9
Electronic density, 153
Electronegativity, 138, 259
  correction for, 145, 148
  global atomic, 262
  orbital, 262
  Pauling, 261
Enantioselective reactions
  hydrogenation, 324
  hydrosilylation, 324
Engelhardt-Radeglia-Wolff (ERW) shielding theory, 134, 155
Ethylchlorosilanes, 3

Fillers of polymers, 11, 17
Functional group, 1, 4

Gated decoupling, 124
Ground state O-Si coordination, 105
Group I derivatives, NMR spectra of, 161
Group II derivatives, NMR spectra of, 161
Group IV derivatives, NMR spectra of, 161
Group V derivatives, NMR spectra of, 166
Group VI derivatives, NMR spectra of, 174

Group VII derivatives, NMR spectra of, 184

Hybrid orbitals, 241
Hybridization, 238, 241
  sp^3, sd^3, 242, 244
Hydrogenation, enantioselective, 3, 24
Hydrolysis, 1, 9, 10
Hydrosilylation, enantioselective, 3, 24
Hyperconjugation, 141, 234, 250, 253, 257, 291
  experimental evidence of, 241
  extent of, 247
  HMO model of, 241
  theoretical aspects of, 244, 247

Ionic character, 138
Immobilization of
  cysteine, 22, 23
  enzymes, 18, 21
  lysine, 21
  metal complex catalysts, 2, 25
  pepsin, 21, 22
  peptides, 18, 21
  proteins, 18
  tyrosine, 21, 22
INADEQUATE method, 125, 143
INEPT method, 125
Interaction, bridging, 77, 78
Interaction, intramolecular
  between Si and C, 83, 93, 98, 107
  between Si and N, 85, 94, 100
  between Si and O, 86, 94, 101, 108
  between Si and halogen, 91, 98, 106, 110
Interaction
  $(p-d)\pi$, 141, 190, 208, 219, 233, 249, 250, 252, 264

INDEX

Interaction (continued)
  (p-d)$_\sigma$, 233, 249, 250
Internal rotation, 146
  Fourier component analysis of, 266
Isodesmic reactions, 266, 276

Kaolin, treatment of, 18
Karplus equation, 145

LCAO approximation, 235, 247, 249
LFER, 205, 263
Liquid crystals, 152

Maximum overlap criterion, 243
Merrifield resin, 24
Methylchlorosilanes, 3
Methylsilylene, stability of, 285
Molecular properties
  acidic-basic, 271
  collective, 255
  one electron, 255
NMR, ^{29}Si, 120, 155
  measurements of, 123, 124
  solvent effect, 128

NQR spectra, 261
Nuclear Overhauser effect, 123, 150

One-center calculation, 247, 249
Orbital
  angular part of, 235
  concept of, 234
  Gaussian type, 245
  hybrid, 241
  localized, 255
  radial part of, 235
  Slater type, 136, 244
  strictly localized, 255, 257
Oriented molecules, 151

Paramagnetic shielding, 133

Phenylchlorosilanes, 2
Photoelectron spectra, 251
Polarizability of silicon, 276
Polysilanes, aromatic behavior, 259
PREP method, 125
Propylchlorosilanes, 3
Proton, diastereotopic, 198
PSSPI method, 125

Reactions
  isodesmic, 266, 276
  of organosilicon acetals, hydrolysis, 67
  of organosilicon alcohols, 59
    acid-base equilibria, 70
    acid-catalyzed rearrangement, 60
    addition to carbonyl bond, 59
    cleavage of benzyl-silicon bond by base, 62
    displacement of hydroxyl by halogen, 60
  of organosilicon alkenes, 40
    free-radical additions, 43
    heterolytic additions, 41
  of organosilicon alkyl acetates, 63
    acid hydrolysis, 66
    alkaline hydrolysis, 63
  of organosilicon alkyl halides, 73
    acid-base equilibria, 81
    carbonium ion forming reactions, 74
    Friedel-Crafts alkylation, 81
    hydrolysis of Si-H bond, 80
    reaction with Grignard reagent, 80
    reaction with sodium atoms, 79
    S_N2 displacement reactions, 76

Reactions (continued)
  of organosilicon alkyl sulfonates, acetolysis, 63
  of organosilicon amines, acid-base equilibria, 55
  of organosilicon aromatic compounds, 46
    dissociation of benzoic acids, 50
    electrophilic substitutions, 46
    solvolysis of cumyl chlorides, 52
  of organosilicon carboxylic acids, dissociation, 54
  of organosilicon esters, hydrolysis, 54
  of organosilicon ethers, 67, 68
    acid-base equilibria, 70
    hydrolysis, 67
    methanolysis, 68
  of organosilicon ketones, dissociation of cyanohydrins, 53
  of organosilicon thiols, heterolytic addition, 72
Reference, NMR, 126, 128
Relaxation reagents, 124
Relaxation time, 124
Resonance, 256
Retention time, 9
Rotational potential, analysis of, 267

Shielding
  diamagnetic, 133
  paramagnetic, 133
  theory (ERW), 134
Shielding constant, relative, 137
Shift reagent, 151
Silabenzene, quantum-chemical calculation, 288
Silaethene, geometry of, 285
Silanone, quantum-chemical calculation, 287

Silicenium ion, calculated stability, 276
Silicon
  multiple bonded, 282
  reactivity of, 283
  stability of, 284
  polarizability of, 276
Silicon directed migrations, 60, 61
Silyl anion, stability of, 277
Silyl substituent, electronic effect, 263
Slater orbitals, 136, 244
Solid state synthesis of peptides, 24
Spacers, 25
Spectra
  charge-transfer, 241
  NQR, 251
  photoelectron, 232
  of silaethene, 285
  UV, 253
SPI method, 125
SPT method, 125
Stability
  increased
    of α-silyl anions, 277
    of α-silylated compounds, 39
    of β-silylated compounds, 40
  of multiple bonded silicon, 284
  of silyl anion, 277
Stabilization
  of carbanion, 40
  of carbonium cation, 62
  of oxonium ion, 67
  of radical, 43, 44
Stationary phases, 9
Steric hindrance
  to aromatic ortho substitution, 50
  of coordination site, 2
  to nucleophilic displacement at silicon, 39

Steric hindrance (continued)
  of reaction center, 63, 64
  of solvation, 56, 64
Substituent chemical shift,
    SCS, 140, 154
Substituent classification,
    129
Substituent constants, 203
  of silyl groups, 51, 69, 71,
    80, 81, 215, 264
Substituent effect, 39, 203
  additivity of, 153, 218
  transmission, 208
Surface tension, 11
Surface treatment, 10, 17
Symmetry, 249

Tensile strength, 18
Thermal expansion
  of glasses, 11
  of polymers, 11
Thermosets, 16
Thermostability, 9, 14
Thixotropy, 11
Trisyl group, 39, 40

UV spectra, 253

Valence shell expansion, 233,
    243
Valence states
  sd^3, 242, 244
  sp^3, 242, 244, 262

Wettability, 11

THE LIBRARY
ST. MARY'S COLLEGE OF MARYLAND
ST. MARY'S CITY, MARYLAND 20686